Lecture Notes in Computer Science   14826

Founding Editors

Gerhard Goos
Juris Hartmanis

W0225711

Editorial Board Members

The series Lecture Notes in Computer Science (LNCS), including its subseries Lecture Notes in Artificial Intelligence (LNAI) and Lecture Notes in Bioinformatics (LNBI), has established itself as a medium for the publication of new developments in computer science and information technology research, teaching, and education.

LNCS enjoys close cooperation with the computer science R & D community, the series counts many renowned academics among its volume editors and paper authors, and collaborates with prestigious societies. Its mission is to serve this international community by providing an invaluable service, mainly focused on the publication of conference and workshop proceedings and postproceedings. LNCS commenced publication in 1973.

Arnaud Devos · András Horváth · Sabina Rossi
Editors

Analytical and Stochastic Modelling Techniques and Applications

28th International Conference, ASMTA 2024
Venice, Italy, June 14, 2024
Proceedings

 Springer

Editors
Arnaud Devos (iD)
Ghent University
Ghent, Belgium

András Horváth (iD)
University of Turin
Turin, Italy

Sabina Rossi (iD)
Università Ca' Foscari Venezia
Venice, Italy

ISSN 0302-9743   ISSN 1611-3349 (electronic)
Lecture Notes in Computer Science
ISBN 978-3-031-70752-0   ISBN 978-3-031-70753-7 (eBook)
https://doi.org/10.1007/978-3-031-70753-7

Preface

It is our great pleasure to present the Proceedings of the 28th International Conference on Analytical & Stochastic Modeling Techniques & Applications (ASMTA 2024). This year's conference continued the tradition of excellence in performance evaluation and stochastic modeling by presenting the latest advances and applications in these dynamic research areas.

The ASMTA 2024 conference was held as a workshop within the prestigious ACM SIGMETRICS/IFIP Performance 2024 conference, which was held in Venice, Italy, June 14, 2024. This provided a unique platform for collaboration and exchange of ideas between researchers and practitioners. We are delighted to be part of this larger event, which promises to foster cross-disciplinary interactions and inspire new research directions.

This volume represents the culmination of rigorous discussion, revision, and review of numerous submissions from the global research community. Each submission underwent a single-blind review by exactly three Program Committee members, ensuring the high quality and relevance of the accepted papers. Strict conflict of interest policies were followed for papers co-authored by any program committee member, including the program co-chairs. Committee members who declared a conflict of interest with a paper were recused from the evaluation and discussion of that paper. In total, ASMTA 2024 received 14 submissions, of which 10 were accepted for presentation. These full papers cover a wide range of topics in analytical and stochastic modeling techniques and their applications.

We were honored to present a keynote talk by Mirco Tribastone. His talk provided an overview of past and current efforts to address software performance issues using modeling and optimization techniques, including recent initiatives to apply research results in industry. This keynote was a joint effort between ASMTA and the European Performance Engineering Workshop (EPEW).

We hope that the proceedings of ASMTA 2024 will inspire further research and innovation in analytical and stochastic modeling techniques and applications. We look forward to the exciting developments that will emerge from this year's conference and beyond.

June 2024

András Horváth
Arnaud Devos
Sabina Rossi

Organization

General Chair

Sabina Rossi   Università Ca' Foscari Venezia, Italy

Program Committee Chairs

András Horváth   University of Turin, Italy
Arnaud Devos   Ghent University, Belgium

Program Committee

Nail Akar	Bilkent University, Turkey
Elene Anton	Université de Pau et des Pays de l'Adour, France
Konstantin Avrachenkov	Inria, France
Dieter Claeys	Ghent University, Belgium
Ioannis Dimitriou	University of Ioannina, Greece
Antonis Economou	University of Athens, Greece
Dieter Fiems	Ghent University, Belgium
Marco Gribaudo	Politecnico di Milano, Italy
Yezekael Hayel	University of Avignon, France
William Knottenbelt	Imperial College London, UK
Lasse Leskelä	Aalto University, Finland
Martin Lopez Garcia	University of Leeds, UK
Jose Nino-Mora	Carlos III University of Madrid, Spain
Juan F. Perez	Universidad del Rosario, Colombia
Tuan Phung-Duc	University of Tsukuba, Japan
Liron Ravner	University of Haifa, Israel
Marie-Ange Remiche	University of Namur, Belgium
Jacques Resing	Eindhoven University of Technology, The Netherlands

Bruno Sericola   Inria, France
Miklos Telek   Budapest University of Technology and
                            Economics, Hungary
Joris Walraevens   Ghent University, Belgium
Jinting Wang   Central University of Finance and Economics,
                            China

Contents

Markov Chain Aggregation with Error Bounds on Transient Distributions

Fabian Michel$^{(\boxtimes)}$ and Markus Siegle

Universität der Bundeswehr München, Werner-Heisenberg-Weg 39,
85579 Neubiberg, Germany
{fabian.michel,markus.siegle}@unibw.de

Abstract. We extend the existing theory of formal error bounds for the transient distribution of an aggregated (or lumped) Markov chain when compared to the transient distribution of the original chain, for both discrete- and continuous-time Markov chains. In the discrete-time setting, we bound the stepwise increment of the error, and in the continuous-time setting, we bound the rate at which the error grows. We then compare these error bounds with relevant concepts in the literature such as exact and ordinary lumpability as well as deflatability and aggregatability. These concepts define stricter than necessary conditions to identify settings in which the aggregation error is zero. We also consider possible algorithms for finding suitable aggregations for which the formal error bounds are low, and we analyse first experiments with these algorithms on different models.

Keywords: Markov chains · Aggregation · Formal error bounds

1 Introduction

State aggregation in dynamic systems has been studied extensively since the 1960s (see [13]). Due to the curse of dimensionality, models with large state spaces are often computationally intractable without state space reduction, and one basic reduction technique is to aggregate multiple states into a single state in the aggregated model. Conditions under which an aggregated Markov chain is again a Markov chain are well known (see strong/weak lumpability in [8,12]), and various cases where exact transient or stationary probabilities of the original model can be derived from an aggregated model have been analysed (see, e.g. [3]).

However, formal error bounds for the approximation error when exact aggregation is not possible have only been studied rarely. [1] has presented error bounds for the transient distributions of discrete-time Markov chains, derived from an aggregated model. We extend the theory developed in [1] to support a more general way of disaggregation and to the continuous-time domain without falling back on uniformisation. Subsequently, we analyse the cases where the error bounds are zero, show optimality of the bounds, and compare them with lumpability concepts from [2,3,7,8]. We present two different algorithms, one based on [2] and one based on [4], with the goal to identify an aggregation resulting in low error bounds.

A. Devos et al. (Eds.): ASMTA 2024, LNCS 14826, pp. 1–17, 2025.
https://doi.org/10.1007/978-3-031-70753-7_1

2   Preliminaries

2.1   Aggregation of Markov Chains

We consider time-homogeneous discrete- and continuous-time Markov chains (DTMCs and CTMCs) on the state space $S = \{1, \ldots, n\}$. The dynamics are given by the stochastic transition matrix $P \in \mathbb{R}^{n \times n}$ for DTMCs, where we have $P(r, s) = \mathbb{P}[X_{k+1} = s \mid X_k = r]$ if X_k denotes the state of the DTMC at time k. For CTMCs, the dynamics are defined via the generator matrix $Q \in \mathbb{R}^{n \times n}$, where $Q(r, s)$ is the transition rate from r to s, and $Q(r, r) = -\sum_{s \neq r} Q(r, s)$. Given an initial distribution $p_0 \in \mathbb{R}^n$, the transient distribution of a DTMC (respectively CTMC) is given by $p_k = p_0 P^k$ (respectively $p_t = p_0 e^{Qt}$), if we interpret p_k as a row vector.

An aggregation of the state space S consists of a set $\Omega = \{\Omega_1, \ldots, \Omega_m\}$ of m aggregates, where Ω is a partition of S, i.e. $\rho \in \Omega$ is a subset of S which represents all states belonging to one aggregate. The aggregation function $\omega : S \to \Omega$ maps a state s to the aggregate to which s belongs, i.e. $s \in \omega(s)$. We approximate the dynamics of the original Markov chain by defining a stochastic transition matrix $\Pi \in \mathbb{R}^{m \times m}$ for DTMCs and a generator matrix $\Theta \in \mathbb{R}^{m \times m}$ for CTMCs on the aggregated state space. $\Pi(\rho, \sigma)$ for $\rho, \sigma \in \Omega$ should be an approximation of the probability to transition from one aggregate state into another, that is, an approximation of $\mathbb{P}[X_{k+1} \in \sigma \mid X_k \in \rho]$. Note that this probability may now depend on k, in contrast to the probability $\mathbb{P}[X_{k+1} = s \mid X_k = r]$ for $r, s \in S$, which is time-independent. However, we consider only time-independent approximations $\Pi(\rho, \sigma)$. Similarly, for CTMCs, $\Theta(\rho, \sigma)$ should approximate the transition rate from aggregate ρ to aggregate σ.

The aggregation can also be applied to initial and transient distributions. We call $\pi_0 \in \mathbb{R}^m$, defined via $\pi_0(\sigma) = \sum_{s \in \sigma} p_0(s)$ for $\sigma \in \Omega$, the aggregated initial distribution and define aggregated transient distributions via $\pi_k = \pi_0 \Pi^k$ (discrete time) and $\pi_t = \pi_0 e^{\Theta t}$ (continuous time). In order to obtain an approximation of the transient probability for a given state in the original chain, we consider distributions $\alpha_\sigma \in \mathbb{R}^n$ with support on $\sigma \in \Omega$. As a shorthand, we write $\alpha(s) := \alpha_{\omega(s)}(s)$. The value $\alpha(s)$ should approximate the conditional probability of being in state s when we know that we are in aggregate $\omega(s)$, i.e. the probability $\mathbb{P}[X_k = s \mid X_k \in \omega(s)]$. Again, this probability is in general dependent on time, but we only consider time-independent approximations α. We then arrive at an approximated transient distribution $\widetilde{p}_k \in \mathbb{R}^n$ by the following formula: $\widetilde{p}_k(s) = \alpha(s)\pi_k(\omega(s))$, or, for continuous time, $\widetilde{p}_t(s) = \alpha(s)\pi_t(\omega(s))$.

We further define the aggregation matrix Λ and the disaggregation matrix A as follows:

$$\Lambda = \begin{pmatrix} | & & | \\ \mathbb{1}_{\Omega_1} & \cdots & \mathbb{1}_{\Omega_m} \\ | & & | \end{pmatrix} \in \mathbb{R}^{n \times m}, \quad A = \begin{pmatrix} -\alpha_{\Omega_1}- \\ \vdots \\ -\alpha_{\Omega_m}- \end{pmatrix} \in \mathbb{R}^{m \times n} \qquad \text{(note: } A\Lambda = I\text{)}$$

We set $\widetilde{P} = \Lambda\Pi A$ and $\widetilde{Q} = \Lambda\Theta A$. This implies in particular that

$$\widetilde{p}_k \widetilde{P}^l = \underbrace{\widetilde{p}_k \Lambda}_{\pi_k} \Pi^l A = \pi_{k+l} A = \widetilde{p}_{k+l}$$

$$\widetilde{p}_t e^{\widetilde{Q}u} = \widetilde{p}_t \sum_{k=0}^{\infty} \frac{u^k}{k!} \widetilde{Q}^k = \underbrace{\widetilde{p}_t \Lambda}_{\pi_t} \left(\sum_{k=0}^{\infty} \frac{u^k}{k!} \Theta^k \right) A = \pi_t e^{\Theta u} A = \pi_{t+u} A = \widetilde{p}_{t+u}$$

$\widetilde{P}$ (which is stochastic) describes the approximated dynamics of the aggregated chain if we blow it up again to the original state space S. It holds that $\widetilde{P}(r,s) = \alpha(s)\Pi(\omega(r),\omega(s))$, so we approximate $P(r,s)$ by the transition probability from aggregate $\omega(r)$ to aggregate $\omega(s)$, weighted with the conditional probability $\alpha(s)$ of being in state s within aggregate $\omega(s)$. We also have $\widetilde{Q}(r,s) = \alpha(s)\Theta(\omega(r),\omega(s))$ and $\widetilde{Q}$ describes the approximated dynamics in a sense, via the equation $\widetilde{p}_t e^{\widetilde{Q}u} = \widetilde{p}_{t+u}$. However, $\widetilde{Q}$ is no longer a generator matrix. The row sums of $\widetilde{Q}$ are 0, but the negative entries are no longer confined to the diagonal.

2.2   Exact Aggregation

Definition 1. *Given a partition Ω of the state space of a DTMC or CTMC, we call a probability distribution p on the state space S **compatible** with distributions α_σ with support on $\sigma \in \Omega$ if $p\Lambda A = p$.*

Compatibility of p and the distributions α means that

$$\alpha(s) = \frac{p(s)}{\sum_{s' \in \omega(s)} p(s')} \text{ for all } s \in S \text{ s.t. } \sum_{s' \in \omega(s)} p(s') > 0$$

Definition 2. *We call an aggregation Ω of the state space of a DTMC (respectively CTMC) with distributions $\alpha_{\Omega_1}, \ldots, \alpha_{\Omega_m}$ and aggregated transition matrix Π (respectively Θ) **dynamic-exact** if $\Pi A = AP$ (respectively $\Theta A = AQ$).*

*If the initial distribution p_0 is further compatible with the α distributions, i.e. if $\widetilde{p}_0 = p_0 \Lambda A = p_0$, then we call the aggregation **exact**.*

If Ω, $\alpha_{\Omega_1}, \ldots, \alpha_{\Omega_m}$ and Π are an exact aggregation, then $\widetilde{p}_k = p_k$ for all k; if the aggregation is dynamic-exact, then it only holds that $\widetilde{p}_k = \widetilde{p}_0 P^k$ (this follows later from Corollary 1). Note the difference to $\widetilde{p}_k = \widetilde{p}_0 \widetilde{P}^k$, which is always true.

The condition $\Pi A = AP$ has appeared in the literature before. Equation (4) on page 135 of [8] states that, if Π is set as in (1) below, and if the α distributions are compatible with the initial distribution p_0, then $\Pi A = AP$ implies weak lumpability of the DTMC. A DTMC is called weakly lumpable for Ω if there exists an initial distribution p_0 such that the process Y_k, defined by

$Y_k = \sigma \in \Omega \iff X_k \in \sigma$, is a Markov chain. For such an initial distribution, the probabilities $\pi_k(\sigma)$ are equal to $\mathbb{P}[Y_k = \sigma] = \mathbb{P}[X_k \in \sigma]$. However, the concept of weak lumpability makes no statement about whether the probability $\mathbb{P}[X_k = s]$ for $s \in \sigma$ can be accurately derived from the knowledge of $\mathbb{P}[X_k \in \sigma]$.

[9, Definition 2.2] defined P to be A-lumpable if $\Pi A = AP$ and if Π is set as in (1). [9] then noted that, given π_k, an exact recovery of $p_k(s)$ is possible if the initial distribution p_0 is compatible with the α distributions. An exact aggregation is also called backward bisimulation of type 2 in [5, Definition 4.3].

2.3   Aggregated Dynamics

Choosing the Aggregates. We consider different possibilities for choosing the aggregates later. For now, assume Ω is fixed. We want to choose Π (or Θ) and α in a way which results in a good approximation of the original dynamics.

Aggregated Transition and Generator Matrices. We consider the following aggregated transition probabilities and rates for $\rho, \sigma \in \Omega$:

$$\Pi(\rho, \sigma) = \sum_{r \in \rho} \alpha(r) \sum_{s \in \sigma} P(r, s) \qquad \Theta(\rho, \sigma) = \sum_{r \in \rho} \alpha(r) \sum_{s \in \sigma} Q(r, s) \qquad (1)$$

The probability $\mathbb{P}[X_{k+1} \in \sigma \mid X_k \in \rho]$ is approximated via the weighted average (with weights $\alpha(r)$) of the probability to transition from a single state $r \in \rho$ into any of the states in σ, since we assume that if we are in aggregate ρ, the probability to be in state $r \in \rho$ is given (approximately) by $\alpha(r)$. In matrix notation, setting Π and Θ as in (1) corresponds to $\Pi = AP\Lambda$ and $\Theta = AQ\Lambda$. Note that Π is again stochastic, and Θ is again a generator matrix. Different choices for Π are discussed in [1], and the above choice yields good approximations in terms of the transient distribution in the experiments done in [1]. Only a so-called "median-based scheme" (see [1, equation (21)]) performs better in some settings. However, this scheme might result in a non-stochastic Π such that the aggregated chain can no longer be considered as a Markov chain.

Conditional Distributions. We also need to choose the conditional distributions α_σ. For DTMCs, the following definitions provided good results, and are compatible with the aggregation techniques which will be considered later.

– The first possibility, called **proportional** α, is given by

$$\alpha(s) = \frac{\sum_{r \in S} P(r, s)}{\sum_{r \in S} \sum_{s' \in \omega(s)} P(r, s')} \qquad (2)$$

$\alpha(s)$ is the same as the probability of being in state s, conditioned on being in the aggregate of s, after the Markov chain took a single step, starting with a uniform distribution. Intuitively, the distributions α_σ should be approximations of this type of conditional probabilities, with the exception that we do not necessarily start with a uniform distribution. Note: proportional α is well-defined if the chain is irreducible.

– The second possibility, called **uniform** α, is given by $\alpha(s) = \frac{1}{|\omega(s)|}$.

For CTMCs, (2) cannot be used since the sum $\sum_{r \in S} Q(r, s)$ could be negative. We therefore use the following for CTMCs:

$$\alpha(s) = \left(\sum_{r \in S,\, r \notin \omega(s)} Q(r, s) \right) \left(\sum_{r \in S,\, r \notin \omega(s)} \sum_{s' \in \omega(s)} Q(r, s') \right)^{-1} \tag{3}$$

3 Bounding the Approximation Error

3.1 Error Bounds for DTMCs

We follow [1] to derive error bounds for the difference between the transient distribution p_k of the DTMC and the approximation $\widetilde{p}_k$. Assume Ω, Π, and the distributions α_σ are given. We do not use the particular forms of Π and α given in (1) and (2); arbitrary choices are possible. [1] set $\alpha(s) = \frac{1}{|\omega(s)|}$ implicitly. We demonstrate that other choices of α do not significantly change the error bounds derived in [1]. Call $e_k = \widetilde{p}_k - p_k$ the error after step k. We want to bound $\|e_k\|_1 = \sum_{s=1}^n |e_k(s)|$ where $e_k(s)$ is the s-th entry of $e_k \in \mathbb{R}^n$. Note that

$$e_k = \widetilde{p}_{k-1} \cdot \left(\widetilde{P} - P + P \right) - p_{k-1} \cdot P = \widetilde{p}_{k-1} \cdot \left(\widetilde{P} - P \right) + \underbrace{\left(\widetilde{p}_{k-1} - p_{k-1} \right)}_{e_{k-1}} \cdot P$$

$$\implies \|e_k\|_1 \leq \left\| \widetilde{p}_{k-1} \cdot \left(\widetilde{P} - P \right) \right\|_1 + \|e_{k-1} \cdot P\|_1 \tag{4}$$

Lemma 1. *Let $e \in \mathbb{R}^k$ be an arbitrary (row) vector and $P \in \mathbb{R}^{k \times k}$ be an arbitrary stochastic matrix. Then $\|e \cdot P\|_1 \leq \|e\|_1$.*

We omit the simple proof. As a consequence of Lemma 1 and (4), we have

$$\|e_k\|_1 \leq \underbrace{\|e_{k-1}\|_1}_{\text{previous error}} + \underbrace{\left\| \widetilde{p}_{k-1} \cdot \left(\widetilde{P} - P \right) \right\|_1}_{\text{error from using approximated transition probabilities}}$$

We bound the second term as follows (see [1, pages 15-17]):

$$
\left\| \tilde{p}_{k-1} \cdot \left(\tilde{P} - P \right) \right\|_1 = \sum_{\sigma \in \Omega} \sum_{s \in \sigma} \left| \sum_{\rho \in \Omega} \sum_{r \in \rho} \tilde{p}_{k-1}(r) \cdot \left(\tilde{P}(r,s) - P(r,s) \right) \right|
$$

$$
= \sum_{\sigma \in \Omega} \sum_{s \in \sigma} \left| \sum_{\rho \in \Omega} \sum_{r \in \rho} \alpha(r) \pi_{k-1}(\rho) \cdot \left(\alpha(s) \Pi(\rho,\sigma) - P(r,s) \right) \right|
$$

$$
= \sum_{\sigma \in \Omega} \sum_{s \in \sigma} \left| \sum_{\rho \in \Omega} \pi_{k-1}(\rho) \cdot \left(\alpha(s) \Pi(\rho,\sigma) \sum_{r \in \rho} \alpha(r) - \sum_{r \in \rho} \alpha(r) P(r,s) \right) \right|
$$

$$
= \sum_{\sigma \in \Omega} \sum_{s \in \sigma} \left| \sum_{\rho \in \Omega} \pi_{k-1}(\rho) \cdot \left(\alpha(s) \Pi(\rho,\sigma) - \sum_{r \in \rho} \alpha(r) P(r,s) \right) \right| \tag{5}
$$

$$
\leq \sum_{\sigma \in \Omega} \sum_{s \in \sigma} \sum_{\rho \in \Omega} \pi_{k-1}(\rho) \cdot \left| \alpha(s) \Pi(\rho,\sigma) - \sum_{r \in \rho} \alpha(r) P(r,s) \right|
$$

$$
= \sum_{\rho \in \Omega} \pi_{k-1}(\rho) \cdot \underbrace{\sum_{\sigma \in \Omega} \sum_{s \in \sigma} \left| \alpha(s) \Pi(\rho,\sigma) - \sum_{r \in \rho} \alpha(r) P(r,s) \right|}_{=: \, \tau(\rho)} =: \langle \pi_{k-1}, \tau \rangle
$$

In particular, it follows that

$$
\|e_k\|_1 \leq \|e_0\|_1 + \sum_{i=0}^{k-1} \underbrace{\sum_{\rho \in \Omega} \pi_i(\rho) \cdot \tau(\rho)}_{\leq \max_{\rho \in \Omega} \tau(\rho)} \leq \|e_0\|_1 + k \cdot \max_{\rho \in \Omega} \tau(\rho) \tag{6}
$$

Calculating $\tau(\rho)$ for every $\rho \in \Omega$ once thus allows us to bound $\|e_k\|_1$ using just the error at time 0 and the aggregated transition probabilities π_i at times $i = 0, \ldots, k-1$. Even simpler, $\max_{\rho \in \Omega} \tau(\rho)$ gives an upper bound for the error growth in every step which gives an instant linear error bound on $\|e_k\|_1$.

3.2   Error Bounds for CTMCs

We extend the above setting to continuous-time Markov chains. Set

$$
\tau(\rho) := \sum_{\sigma \in \Omega} \underbrace{\sum_{s \in \sigma} \left| \alpha(s) \Theta(\rho,\sigma) - \sum_{r \in \rho} \alpha(r) Q(r,s) \right|}_{=: \, \tau(\rho,\sigma)} \tag{7}
$$

which exactly matches the definition in the case of discrete time. [1] already proved the following: set Θ as in (1) and consider a uniformisation of the original

CTMC with uniformisation rate q. Apply the same aggregation to the resulting DTMC and set Π as in (1). Then, the error growth in step $k \to k+1$ in the DTMC can be bounded by $\frac{1}{q} \cdot \sum_{\rho \in \Omega} \pi_k(\rho) \cdot \tau(\rho)$ where τ is derived for the aggregated CTMC as in (7). We now drop the detour via the uniformisation and claim that the factors τ can be interpreted as a rate of error growth for CTMCs.

Theorem 1. *Let e_t be the vector of component-wise error of the approximated transient distribution at time t, i.e. $e_t = \widetilde{p}_t - p_t$. Then:*

(i) We have the following bound for the error at time t:

$$\|e_t\|_1 \leq \|e_0\|_1 + \int_0^t \langle \pi_s, \tau \rangle \, \mathrm{d}s \qquad where \ \langle \pi_s, \tau \rangle = \sum_{\rho \in \Omega} \pi_s(\rho) \cdot \tau(\rho)$$

and $\|e_t\|_1 \leq \|e_0\|_1 + t \cdot \max_{\rho \in \Omega} \tau(\rho)$

(ii) $\|e_t\|_1$ is absolutely continuous, almost everywhere (a.e.) differentiable, and

$$\frac{\mathrm{d}}{\mathrm{d}t} \|e_t\|_1 \leq \langle \pi_t, \tau \rangle \ a.e. \quad and \quad \limsup_{u \to 0} \frac{\|e_{t+u}\|_1 - \|e_t\|_1}{u} \leq \langle \pi_t, \tau \rangle \ \forall t \geq 0$$

Before being able to prove this, we need another lemma.

Lemma 2. *Assume that $f : \mathbb{R} \to \mathbb{R}$ is differentiable in 0, and that $f(0) = 0$.*

$$Then: \quad \limsup_{u \to 0} \frac{|f(u)|}{u} = \lim_{\substack{u \to 0 \\ u > 0}} \frac{|f(u)|}{u} = |f'(0)|$$

We omit the simple proof and proceed to show Theorem 1.

Proof (of Theorem 1). As (ii) immediately implies (i), it suffices to prove (ii).

First, note the following: every component of e_t is continuously differentiable in t, as both p_t and $\widetilde{p}_t$ are continuously differentiable with respect to t. Indeed, calculating the derivative of all components of e_t simultaneously, we get:

$$\frac{\mathrm{d}}{\mathrm{d}t}(\widetilde{p}_t - p_t) = \frac{\mathrm{d}}{\mathrm{d}t}\left(\widetilde{p}_0 e^{\widetilde{Q}t} - p_0 e^{Qt}\right) = \widetilde{p}_0 e^{\widetilde{Q}t}\widetilde{Q} - p_0 e^{Qt}Q = \widetilde{p}_t\widetilde{Q} - p_t Q \quad (8)$$

$$\left\|\frac{\mathrm{d}}{\mathrm{d}t}(\widetilde{p}_t - p_t)\right\|_1 \leq \left\|\widetilde{p}_t\widetilde{Q}\right\|_1 + \|p_t Q\|_1 \leq |S| \cdot \left(\max_{r,s \in S}\left|\widetilde{Q}(r,s)\right| + \max_{r,s \in S}|Q(r,s)|\right)$$

As every component of e_t is continuously differentiable with bounded derivative, $\|e_t\|_1$ is absolutely continuous and differentiable a. e. (see [11, Section 5.4]).

$$\|e_{t+u}\|_1 = \|\widetilde{p}_{t+u} - p_{t+u}\|_1 = \left\|\widetilde{p}_t\left(e^{\widetilde{Q}u} - e^{Qu}\right) + (\widetilde{p}_t - p_t)e^{Qu}\right\|_1$$

$$\leq \left\|\widetilde{p}_t\left(e^{\widetilde{Q}u} - e^{Qu}\right)\right\|_1 + \left\|(\widetilde{p}_t - p_t)e^{Qu}\right\|_1$$

$$\overset{\textcircled{\tiny ©}}{\leq} \left\|\widetilde{p}_t\left(e^{\widetilde{Q}u} - e^{Qu}\right)\right\|_1 + \|\widetilde{p}_t - p_t\|_1 = \left\|\widetilde{p}_t\left(e^{\widetilde{Q}u} - e^{Qu}\right)\right\|_1 + \|e_t\|_1$$

$$\implies \|e_{t+u}\|_1 - \|e_t\|_1 \leq \left\|\widetilde{p}_t\left(e^{\widetilde{Q}u} - e^{Qu}\right)\right\|_1 = \sum_{s \in S}\left|\left(\widetilde{p}_t\left(e^{\widetilde{Q}u} - e^{Qu}\right)\right)(s)\right| \quad (9)$$

where ⊙ follows from Lemma 1 since e^{Qu} is a stochastic matrix. We now take a closer look at the right hand side. In particular, we are interested in

$$\frac{\mathrm{d}}{\mathrm{d}u}\widetilde{p}_t\left(e^{\widetilde{Q}u}-e^{Qu}\right)=\widetilde{p}_t\left(e^{\widetilde{Q}u}\widetilde{Q}-e^{Qu}Q\right)\overset{\text{for }u=0}{=}\widetilde{p}_t\left(\widetilde{Q}-Q\right)$$

$$\implies\left.\frac{\mathrm{d}}{\mathrm{d}u}\right|_{u=0}\left(\widetilde{p}_t\left(e^{\widetilde{Q}u}-e^{Qu}\right)\right)(s)=\sum_{r\in S}\widetilde{p}_t(r)\left(\widetilde{Q}(r,s)-Q(r,s)\right)$$

Hence

$$\limsup_{u\to0}\frac{\|e_{t+u}\|_1-\|e_t\|_1}{u}\overset{(9)}{\le}\sum_{s\in S}\limsup_{u\to0}\frac{\left|\left(\widetilde{p}_t\left(e^{\widetilde{Q}u}-e^{Qu}\right)\right)(s)\right|}{u}$$

$$\overset{\text{Lemma 2}}{=}\sum_{s\in S}\left|\sum_{r\in S}\widetilde{p}_t(r)\left(\widetilde{Q}(r,s)-Q(r,s)\right)\right|\overset{\circledast}{\le}\sum_{\rho\in\Omega}\pi_t(\rho)\cdot\tau(\rho)=\langle\pi_t,\tau\rangle$$

where ⊛ follows from the same calculation as for DTMCs, see (5). □

3.3   When is the Error Bound 0?

In order to better understand the error bound which arises from the τ factors, this section first analyses in which cases the error bound is equal to 0, and then shows tightness of the error bound.

Lemma 3. *Given a DTMC, a partition Ω of its state space, arbitrary distributions α_σ with support on $\sigma\in\Omega$, and arbitrary Π, it holds that*

$$\forall\rho\in\Omega:\tau(\rho)=0\iff\Pi A=AP\overset{\text{Definition 2}}{\iff}\text{the aggregation is dynamic-exact}$$

The same holds for CTMCs with Π replaced by Θ and P replaced by Q.

Proof. Note: $\tau(\rho)=0$ if, and only if, $\tau(\rho,\sigma)=0$ (defined in (7)) for all σ. Hence,

$$\forall\rho\in\Omega:\tau(\rho)=0\iff\forall\rho,\sigma\in\Omega:\tau(\rho,\sigma)=0$$

$$\iff\forall\rho,\sigma\in\Omega:\sum_{s\in\sigma}\left|\alpha(s)\Pi(\rho,\sigma)-\sum_{r\in\rho}\alpha(r)P(r,s)\right|=0 \tag{10}$$

$$\iff\forall\rho,\sigma\in\Omega:\forall s\in\sigma:\;\alpha(s)\Pi(\rho,\sigma)=\sum_{r\in\rho}\alpha(r)P(r,s)$$

This already proves Lemma 3 since

$$\alpha(s)\Pi(\rho,\sigma)=\underbrace{(\Pi A)}_{\in\mathbb{R}^{m\times n}}(\rho,s)\qquad\text{and}\qquad\sum_{r\in\rho}\alpha(r)P(r,s)=(AP)(\rho,s)$$

The same calculation is true with Π replaced by Θ and P replaced by Q. Also note that one can even show that $\max_{\rho\in\Omega}\tau(\rho)=\|\Pi A-AP\|_\infty$. □

Corollary 1. *Given a DTMC or CTMC, a partition Ω, arbitrary distributions α_σ with support on $\sigma \in \Omega$, and arbitrary Π (respectively Θ) such that $\Pi A = AP$ (respectively $\Theta A = AQ$, i.e. a dynamic-exact aggregation), it holds that*

$$\|\widetilde{p}_k - p_k\|_1 \leq \|\widetilde{p}_0 - p_0\|_1 \quad \text{or, for continuous time,} \quad \|\widetilde{p}_t - p_t\|_1 \leq \|\widetilde{p}_0 - p_0\|_1$$

In particular, if an aggregation is exact, then $\|\widetilde{p}_k - p_k\|_1 = 0$ for all k (respectively $\|\widetilde{p}_t - p_t\|_1 = 0$ for all t).

This follows from Lemma 3 and (6) (respectively Theorem 1 for CTMCs). We next show tightness of the error bounds.

Theorem 2. *Given a DTMC or CTMC, a partition Ω, distributions α_σ with support on $\sigma \in \Omega$, and arbitrary Π or Θ, assume that $\tau(\rho) > 0$ for some ρ. Then, there exists an initial distribution p_0 which is compatible with the α distributions such that $\|\widetilde{p}_1 - p_1\|_1 = \langle \pi_0, \tau \rangle$ or, for CTMCs, $\lim_{t \to 0, t > 0} \frac{1}{t} \|\widetilde{p}_t - p_t\|_1 = \langle \pi_0, \tau \rangle$.*

Proof. There must be some aggregate $\rho \in \Omega$ with $\tau(\rho) > 0$. We choose $p_0 = \alpha_\rho$, which is clearly compatible with the α distributions (hence $\widetilde{p}_0 = p_0 = \alpha_\rho$). In the discrete-time case, we have $p_1 = \alpha_\rho P$ and $\widetilde{p}_1 = \alpha_\rho \Lambda \Pi A$. Hence

$$\|\widetilde{p}_1 - p_1\|_1 = \sum_{s \in S} |\widetilde{p}_1(s) - p_1(s)| = \sum_{s \in S} \left| \alpha(s) \Pi(\rho, \omega(s)) - \sum_{r \in \rho} \alpha(r) P(r, s) \right|$$

$$= \sum_{\sigma \in \Omega} \sum_{s \in \sigma} \left| \alpha(s) \Pi(\rho, \sigma) - \sum_{r \in \rho} \alpha(r) P(r, s) \right| = \tau(\rho) = \langle \pi_0, \tau \rangle$$

since π_0 is the Dirac measure on ρ. We proceed similarly for the continuous-time case, again setting $p_0 = \alpha_\rho$. Note that, as already established in (8), we have

$$\frac{d}{dt}\bigg|_{t=0} (\widetilde{p}_t - p_t)(s) = \left(\widetilde{p}_0 \widetilde{Q} - p_0 Q \right)(s) \overset{\widetilde{p}_0 = p_0 = \alpha_\rho}{=} \left(\alpha_\rho \widetilde{Q} - \alpha_\rho Q \right)(s)$$

$$= \left(\alpha_\rho \Lambda \Theta A - \alpha_\rho Q \right)(s)$$

Hence, noting that $\|\widetilde{p}_0 - p_0\|_1 = 0$, we obtain

$$\lim_{\substack{t \to 0 \\ t > 0}} \frac{1}{t} \|\widetilde{p}_t - p_t\|_1 = \sum_{s \in S} \lim_{\substack{t \to 0 \\ t > 0}} \frac{|(\widetilde{p}_t - p_t)(s)|}{t} \overset{\text{Lemma 2}}{=} \sum_{s \in S} |(\alpha_\rho \Lambda \Theta A - \alpha_\rho Q)(s)|$$

$$= \ldots \text{ (as in the discrete-time case)} = \langle \pi_0, \tau \rangle$$

$\square$

4 Lumpability and Aggregatability

The following definition was given in [3, Definition 1]:

Definition 3. *A partition $\Omega = \{\Omega_1, \ldots, \Omega_m\}$ of the state space of a DTMC is called **ordinarily lumpable** if*

$$\forall r, r' \in S \ s.t. \ \omega(r) = \omega(r') : \forall \sigma \in \Omega : \qquad \sum_{s \in \sigma} P(r, s) = \sum_{s \in \sigma} P(r', s) \qquad (11)$$

For CTMCs, Ω is called ordinarily lumpable if (11) holds with P replaced by Q.

For an ordinarily lumpable partition, we have that $\pi_k(\sigma) = \sum_{s \in \sigma} p_k(s)$ if Π is set as in (1), and for any initial distribution p_0 (and independently of the choice of α). See [3, Theorem 5]. The same holds in the continuous-time case. [3, Definition 1] also defines exact lumpability:

Definition 4. *A partition $\Omega = \{\Omega_1, \ldots, \Omega_m\}$ of the state space of a DTMC is called **exactly lumpable** if*

$$\forall s, s' \in S \ s.t. \ \omega(s) = \omega(s') : \forall \rho \in \Omega : \qquad \sum_{r \in \rho} P(r, s) = \sum_{r \in \rho} P(r, s') \qquad (12)$$

*For CTMCs, Ω is called exactly lumpable if (12) holds with P replaced by Q. Ω is further called **strictly lumpable** if it is both ordinarily and exactly lumpable.*

[2, Definition 2.1] also defines lumpability. Note that this definition of lumpability agrees with the definition of ordinary lumpability given above. [2, Definition 2.1] further defines deflatability and aggregatability:

Definition 5. *A partition Ω of the state space of a DTMC, together with distributions $\alpha_\sigma \in \mathbb{R}^n$ with support on $\sigma \in \Omega$, is called **deflatable** if*

$$\forall r \in S : \forall s \in S : \qquad P(r, s) = \alpha(s) \cdot \sum_{s' \in \omega(s)} P(r, s') \qquad (13)$$

*The partition Ω, together with distributions α, is further called **aggregatable** if it is deflatable and if Ω is ordinarily lumpable.*

Note that aggregatability implies that $\widetilde{P} = P$ by [2, Proposition 2.6] (if Π is set as in (1)). Definition 5 cannot be extended to CTMCs easily.

Proposition 1. *Assume an irreducible DTMC or CTMC is exactly lumpable w.r.t. the partition Ω. When setting Π (respectively Θ) as in (1) and using proportional α as in (2) (respectively (3)) or uniform α, then it holds that*

$$\tau(\rho) = 0 \ \forall \rho \in \Omega \qquad and \qquad \alpha(s) = \frac{1}{|\omega(s)|} \ \forall s \in S$$

Proof (Sketch). First, we look at DTMCs. Assume α is set as in (2), and $\omega(s) = \omega(s') = \sigma$. Then:

$$\alpha(s') \cdot \sum_{r \in S} \sum_{s'' \in \sigma} P(r, s'') = \sum_{r \in S} P(r, s') \overset{\circledast}{=} \sum_{r \in S} P(r, s) = \alpha(s) \cdot \underbrace{\sum_{r \in S} \sum_{s'' \in \sigma} P(r, s'')}_{>0 \text{ by irreducibility}}.$$

where $\circledast$ holds by exact lumpability. Hence $\alpha(s) = \alpha(s')$. For CTMCs and α as in (3), the calculation is similar. Now, one way to proceed is to show that $\tau(\rho) = 0$ for all ρ is equivalent to:

$$\forall s, s' \text{ s.t. } \omega(s) = \omega(s') : \forall \rho \in \Omega : \frac{\sum_{r \in \rho} \alpha(r) P(r, s)}{\alpha(s)} = \frac{\sum_{r \in \rho} \alpha(r) P(r, s')}{\alpha(s')} \quad (14)$$

We skip the proof of equivalence here, but it relies only on simple calculations based on (10). More details are given in [10, Proposition 8] or in [7, Section 2 and 3], where the same equivalence is shown in a slightly different context (note that (14) is equal to [7, (Cond1) on page 771]). If we plug $\alpha(s) = \frac{1}{|\omega(s)|}$ into (14), we get the defining equation (12) of exact lumpability, finishing the proof. $\square$

Proposition 2. *Given an irreducible DTMC and a partition Ω, assume*

$$\forall r \in S : \forall s \in S : \qquad P(r, s) = c(s) \cdot \sum_{s' \in \omega(s)} P(r, s')$$

for constants $c(s) \in [0, 1]$ which only depend on $s \in S$ (this is equivalent to the existence of distributions α s.t. Ω and α are deflatable). When setting Π as in (1) and using proportional α as in (2), it holds that

$$\tau(\rho) = 0 \ \forall \rho \in \Omega \qquad and \qquad \alpha(s) = c(s) \ \forall s \in S$$

We omit the proof of this proposition, it relies only on basic calculations and the equivalence of $\tau(\rho) = 0$ for all ρ to (14). We next show that none of the lumpability concepts above are necessary conditions for $\tau(\rho) = 0$ for all ρ. Except for [7], a large part of the literature has thus treated stricter than necessary conditions in order for dynamic-exact aggregation to be possible.

Proposition 3. *There are partitions Ω of the state space of a DTMC and probability distributions α_σ with support on $\sigma \in \Omega$ which are dynamic-exact (when Π is set as in (1)), but where Ω is neither ordinary lumpable, nor exactly lumpable, nor are Ω and the distributions α deflatable.*

Proof. Consider the state space $S = \{1, 2, 3\}$, the aggregation $\Omega = \{\{1\}, \{2, 3\}\}$ and $\alpha(1) = 1, \alpha(2) = \frac{1}{4}, \alpha(3) = \frac{3}{4}$ as well as the DTMC given by:

$$P = \begin{pmatrix} 0 & \frac{1}{4} & \frac{3}{4} \\ 0 & \frac{1}{2} & \frac{1}{2} \\ \frac{4}{9} & \frac{1}{18} & \frac{1}{2} \end{pmatrix} \overset{(1)}{\Longrightarrow} \Pi = \begin{pmatrix} 0 & 1 \\ \frac{1}{3} & \frac{2}{3} \end{pmatrix}, \quad A = \begin{pmatrix} 1 & 0 & 0 \\ 0 & \frac{1}{4} & \frac{3}{4} \end{pmatrix}, \quad \Lambda = \begin{pmatrix} 1 & 0 \\ 0 & 1 \\ 0 & 1 \end{pmatrix}$$

$\Pi A = AP$ holds, i.e. this aggregation is dynamic-exact and $\tau(\{1\}) = \tau(\{2, 3\}) = 0$. We show that none of the stated properties hold for Ω and α:

- ordinary lumpability: since $\omega(2) = \omega(3)$, by (11) in Definition 3, we would need $1 = P(2, 2) + P(2, 3) = P(3, 2) + P(3, 3) = \frac{5}{9}$ which is clearly not true.
- exact lumpability: since $\omega(2) = \omega(3)$, by (12) in Definition 4, we would need $\frac{5}{9} = P(2, 2) + P(3, 2) = P(2, 3) + P(3, 3) = 1$ which is clearly not true.
- deflatability: since $\omega(2) = \omega(3)$, by (13) in Definition 5, we would need

$$\frac{1}{2} = P(2, 2) = \alpha(2) \sum_{s \in \{2,3\}} P(2, s) = \alpha(2) \cdot 1 \implies \alpha(2) = \frac{1}{2}$$

$$\frac{1}{18} = P(3, 2) = \alpha(2) \sum_{s \in \{2,3\}} P(3, s) = \alpha(2) \cdot \frac{5}{9} \implies \alpha(2) = \frac{1}{10}$$

so Ω and α are not deflatable, and there is no other deflatable choice for α.

In fact, there is no aggregation with $\Omega \neq \{S\}$ and $\Omega \neq \{\{1\}, \{2\}, \{3\}\}$ which is ordinary or exactly lumpable, or for which deflatable α distributions exist. $\square$

For this example, neither using proportional α as in (2) nor uniform α delivers the choice of α which results in a dynamic-exact aggregation. We get $\alpha(1) = 1$, $\alpha(2) = \frac{29}{92} \approx 0.315, \alpha(3) = \frac{63}{92} \approx 0.685$ for proportional α. The proposed ways of calculating α can thus only be seen as an approximation of the optimal choice.

5   Choosing the Aggregates

We want to choose a partition Ω such that the error bounds $\tau(\rho)$ are low, as this results in a good approximation $\widetilde{p}_k$ of p_k. In order to reduce the computational effort required to calculate $\widetilde{p}_k$, we would also like $|\Omega| = m \ll n = |S|$. An ideal algorithm would receive a parameter ε as input and determine the partition Ω with the fewest aggregates satisfying $\max_{\rho \in \Omega} \tau(\rho) < \varepsilon$. This would guarantee a stepwise error (or error growth rate) of at most ε (see (6) and Theorem 1 (i)). Solving this problem exactly will in general result in a runtime exceeding the time needed to compute p_k exactly for the original chain. We therefore consider different ways of choosing an Ω which is close to the optimal solution.

5.1   Almost Aggregatability

We consider three algorithms based on [2] which identify almost aggregatable partitions for which the error bounds are low by Proposition 2. These algorithms use the singular value decomposition of P.

- **SVD sgn**: proposed as a simple algorithm in [2] with only limited practical applicability due to its numerical instability.
- **SVD seba**: proposed as a stable variant in [2] via a combination with [6].
- **SVD dir**: a new algorithm devised by us based on SVD sgn. The SVD algorithm from [2] basically assigns a (cropped) row vector of the right hand matrix V of the singular value decomposition $U \Sigma V^\mathsf{T}$ of P to every state in the Markov chain. SVD sgn then only analyses the sign structure of these vectors, while SVD dir exploits the fact that the vectors of two states in the same aggregate should point in approximately the same direction.

The three variants of the SVD algorithm can only be applied to DTMCs (aggregatable was only defined for DTMCs in Definition 5). In order to decide how coarse the aggregation should be, all three algorithms receive a parameter ε as input which is used to decide where to cut off the row vectors of V used for partitioning, i.e. ε is used to decide which dimension these vectors should have. Details are given in [10, Section 5.2].

5.2 ε-Almost Exact Lumpability

By Proposition 1, if Ω is exactly lumpable, Π is set as in (1) and if proportional α as in (2) or uniform α is used, then the error bound is zero. In the general case, it is more likely that a partition exists which is close to being exactly lumpable. This motivates the following definition:

Definition 6. *We call a partition Ω ε-almost exactly lumpable if:*

$$\forall s, s' \in S \ s.t. \ \omega(s) = \omega(s') : \qquad \sum_{\rho \in \Omega} \left| \sum_{r \in \rho} P(r, s) - \sum_{r \in \rho} P(r, s') \right| \leq \varepsilon$$

We now develop an algorithm (see Algorithm 1, similar to the ideas from [4, p. 269]) which finds an ε-almost exactly lumpable partition. It works for DTMCs as well as CTMCs. For a given ε, the algorithm should find a partition which is as coarse as possible and still satisfies ε-almost exact lumpability. The idea of the algorithm is as follows: we start with the initial partition $\Omega = \{S\}$, which is then successively refined. At every refinement step, for every aggregate $\sigma \in \Omega$ and for all states $s \in \sigma$, we construct vectors of incoming probabilities

$$\text{inc}(s) = \left(\sum_{r \in \Omega_1} P(r, s), \ \ldots, \ \sum_{r \in \Omega_m} P(r, s) \right) \in \mathbb{R}^m$$

where m is the current number of aggregates in Ω. By Definition 6, we have that the current partition Ω is ε-almost exactly lumpable if, and only if, we have that $\|\text{inc}(s) - \text{inc}(s')\|_1 \leq \varepsilon$ for all states s and s' belonging to the same aggregate σ. If this is not the case, the algorithm proceeds with the refinement by partitioning the states into smaller aggregates. This procedure stops when an ε-almost exactly lumpable partition is found (at the latest when every aggregate consists of a single state).

Algorithm 1. Calculating almost exactly lumpable partitions

Input: a Markov chain, defined via its transition matrix P on state space S,
and the parameter ε (a generator matrix Q can be used instead of P)
Output: an aggregation function ω
whose corresponding partition is ε-almost exactly lumpable

1: $\omega^{(1)} \leftarrow ((s \in S) \mapsto 1)$ ▷ aggregation function
2: $i \leftarrow 1$ ▷ iteration counter
3: $m \leftarrow 1$ ▷ number of aggregates
4: **repeat**
5:     $m_{\text{old}} \leftarrow m$ ▷ saves number of old aggregates
6:     $m \leftarrow 0$ ▷ counts number of new aggregates
7:     **for all** $j \in \{1, \ldots, m_{\text{old}}\}$ **do**   ▷ loop over old aggregates
8:        **for all** $s \in \{r \in S : \omega^{(i)}(r) = j\}$ **do**   ▷ loop over states in same aggregate
9:          $\text{inc}(s) \leftarrow \mathbf{0} \in \mathbb{R}^{m_{\text{old}}}$
10:          **for all** $k \in \{1, \ldots, m_{\text{old}}\}$ **do**   ▷ loop over potential splitters
11:            $\text{inc}(s)_k \leftarrow \sum_{r \in S : \omega^{(i)}(r) = k} P(r, s)$ ▷ inc. prob. from agg. k to state s
12:          **end for**
13:        **end for**
14:        $C \leftarrow \text{cluster}(\{r \in S : \omega^{(i)}(r) = j\}, \text{inc}, \varepsilon)$ ▷ see below
15:        **for all** $\sigma \in C$ **do**   ▷ loop over clusters
16:          **for all** $s \in \sigma$ **do**
17:            $\omega^{(i+1)}(s) \leftarrow m + 1$ ▷ states in σ are assigned to the same agg.
18:          **end for**
19:          $m \leftarrow m + 1$ ▷ increment aggregate number
20:        **end for**
21:     **end for**
22:     $i \leftarrow i + 1$
23: **until** $m_{\text{old}} = m$ ▷ stop when no aggregates were split
24: **return** $\omega^{(i)}$

$\text{cluster}(T, f, \varepsilon)$ takes a subset of states $T \subseteq S$, a function $f : T \to \mathbb{R}^k$ and $\varepsilon > 0$ as input. The output is a partition C of T such that for any cluster $\sigma \in C$ and any two states $s, s' \in \sigma$, we have that $\|f(s) - f(s')\|_1 \leq \varepsilon$. We use Python and `scipy.cluster.hierarchy.fclusterdata` to calculate the clustering.

5.3 Experiments

We compare the performance of SVD sgn, SVD seba, SVD dir, and Algorithm 1 by comparing the error bounds given by the τ factors resulting from the aggregations returned by the different algorithms – the lower, the better. By default, we calculate the α distributions as in (2), and Π (or Θ) is set as in (1).

In Fig. 1, we consider a setting for which the SVD algorithms were designed. We see that the SVD variants (except SVD seba) perform better than Algorithm 1 for almost aggregatable chains. The higher stability of SVD dir pays off in comparison to SVD sgn: we see a sharp drop in the error bounds around 20 aggregates (the number of aggregates in the almost aggregatable partition). For SVD sgn, the drop is more a gradual decrease. Algorithm 1 does not identify

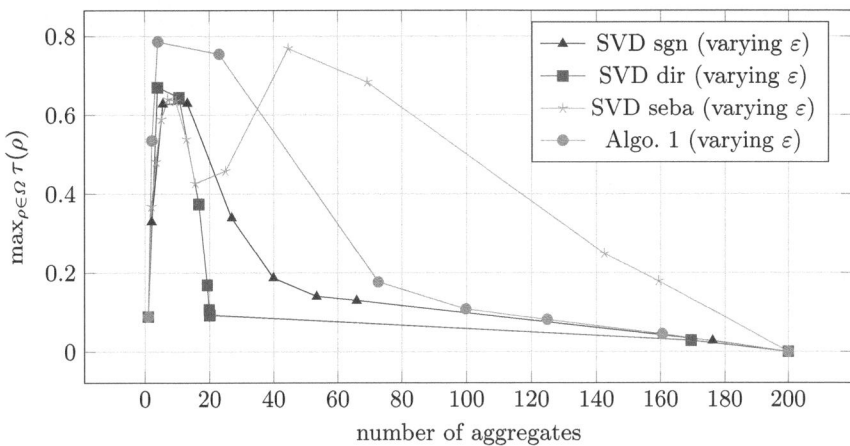

Fig. 1. SVD sgn, SVD dir, SVD seba, and Algorithm 1 executed on 100 randomly generated almost aggregatable DTMCs with 200 states, 20 aggregates and a probability of 0.5 to have no transition between a particular pair of aggregates. The almost aggregatable DTMCs were obtained by random perturbation of the transition matrix of an aggregatable DTMC with a magnitude of 0.002. Each plotted point is an average resulting from running the algorithms with a particular fixed input parameter ε on the 100 DTMCs. Multiple different parameters are used to obtain the different points.

the almost aggregatable partition. SVD seba performs similarly to SVD dir for a low number of aggregates, but there is a sudden change around 20 aggregates when SVD seba starts to perform worse than all other algorithms. This is due to the fact that we limited the maximum number of iterations of the SEBA algorithm (see [6, Algorithm 3.1]) to 300 iterations because of its high runtime. Regardless of the number of maximum iterations, we could never observe SVD seba performing significantly better than SVD dir in all our experiments. The latter is therefore a good alternative. For details on implementation, refer to [10].

We also considered the compositional stochastic process algebra model RSVP from [14], comprising a lower and upper network channel with capacities for M and N calls, and a number of identical mobile nodes which request resources for calls. Due to the identical mobile nodes, lossless aggregation is possible. Comparing the different algorithms in Fig. 2 for a uniformisation of this model, we see that only Algorithm 1 identifies the lossless aggregation (which is exactly lumpable). The SVD variants perform much worse. Figure 2 also compares the two different ways to calculate α: proportional α as in (2) and uniform α.

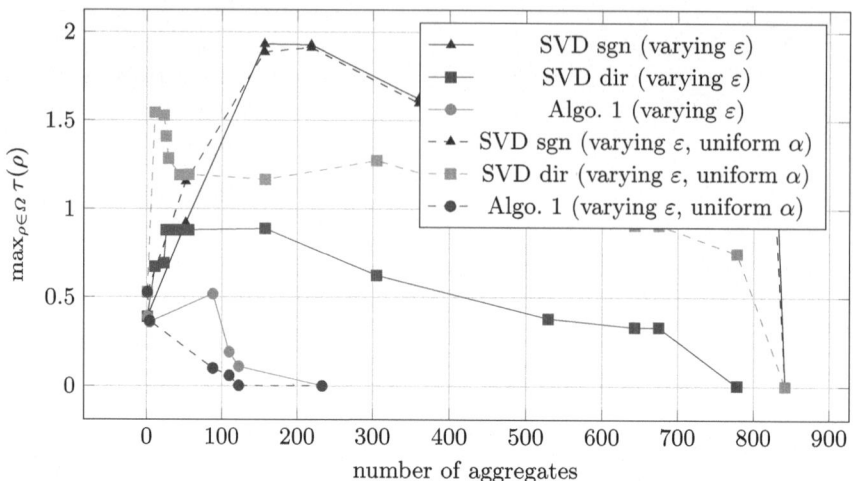

Fig. 2. SVD sgn, SVD dir, and Algorithm 1 executed on the uniformisation of the model from [14] with $M = 7$, $N = 5$ and 3 mobile nodes, resulting in a total of 842 states. By symmetry of the mobile nodes, a lossless reduction to 234 states is possible.

6   Conclusion and Outlook

We extended the error bounds for the difference between the transient distribution of an aggregated chain and the original chain originally derived in [1] to a more general setting and proved that these bounds are tight. We also showed a relation of the error bounds to existing lumpability concepts, and we compared algorithms which identify different settings in which the error bounds are low. To obtain more reliable results, these algorithms have to be compared with other approaches in real-world applications in the future, and a combination with the adaptive aggregation from [1] should be investigated. Another interesting topic would be to develop an efficient algorithm which finds an approximate solution to $\Pi A = AP$, i.e. the most general case in which the error bounds are zero.

References

1. Abate, A., Andriushchenko, R., Češka, M., Kwiatkowska, M.: Adaptive formal approximations of Markov chains. Perform. Eval. **148**, 102207 (2021). https://doi.org/10.1016/j.peva.2021.102207
2. Bittracher, A., Schütte, C.: A probabilistic algorithm for aggregating vastly undersampled large Markov chains. Physica D: Nonlinear Phenom. **416**, 132799 (2021). https://doi.org/10.1016/j.physd.2020.132799
3. Buchholz, P.: Exact and ordinary lumpability in finite Markov chains. J. Appl. Probab. **31**(1), 59–75 (1994). https://doi.org/10.2307/3215235
4. Buchholz, P.: Exact performance equivalence: an equivalence relation for stochastic automata. Theoret. Comput. Sci. **215**, 263–287 (1999). https://doi.org/10.1016/S0304-3975(98)00169-8

5. Buchholz, P.: Bisimulation relations for weighted automata. Theoret. Comput. Sci. **393**, 109–123 (2008). https://doi.org/10.1016/j.tcs.2007.11.018

6. Froyland, G., Rock, C.P., Sakellariou, K.: Sparse eigenbasis approximation: multiple feature extraction across spatiotemporal scales with application to coherent set identification. Commun. Nonlinear Sci. Numer. Simul. **77**, 81–107 (2019). https://doi.org/10.1016/j.cnsns.2019.04.012

7. Ganguly, A., Petrov, T., Koeppl, H.: Markov chain aggregation and its applications to combinatorial reaction networks. J. Math. Biol. **69**(3), 767–797 (2014). https://doi.org/10.1007/s00285-013-0738-7

8. Kemeny, J.G., Snell, J.L.: Finite Markov Chains. Springer, Heidelberg (1976). https://link.springer.com/book/9780387901923

9. Ledoux, J., Truffet, L.: Markovian bounds on functions of finite Markov chains. Adv. Appl. Probab. **33**(2), 505–519 (2001). https://doi.org/10.1017/S0001867800010910

10. Michel, F., Siegle, M.: Markov chain aggregation with error bounds on transient distributions. arXiv preprint (2024). https://doi.org/10.48550/arXiv.2403.07618

11. Royden, H.L.: Real Analysis, 3rd edn. Collier Macmillan (1988)

12. Rubino, G., Sericola, B.: A finite characterization of weak lumpable Markov processes. Part II: the continuous time case. Stoch. Process. Appl. **45**(1), 115–125 (1993). https://doi.org/10.1016/0304-4149(93)90063-A

13. Simon, H.A., Ando, A.: Aggregation of variables in dynamic systems. Econometric **29**(2), 111–138 (1961). https://doi.org/10.2307/1909285

14. Wang, H., Laurenson, D.I., Hillston, J.: Evaluation of RSVP and mobility-aware RSVP using performance evaluation process algebra. In: 2008 IEEE International Conference on Communications, pp. 192–197 (2008). https://doi.org/10.1109/ICC.2008.43

Strong Aggregation in the Stochastic Matching Model with Random Discipline

Jean-Michel Fourneau[1,2(✉)] and Moyi Yang[2]

[1] INRIA ARGO, Paris, France
[2] DAVID, Univ. Paris-Saclay, UVSQ, Versailles, France
Jean-Michel.Fourneau@uvsq.fr, moyi.yang@ens.uvsq.fr

Abstract. We consider a stochastic matching model with a general compatibility graph with self-loops on every node and a random matching policy. We consider the discrete time Markov chain associated with such a model where arrivals of items are independently and identically distributed. Due to the self-loops in the compatibility graph, the states of this chain are exactly the independent sets of the graph. We prove that this chain is ordinary lumpable if the automorphism group of the compatibility graph is non-trivial. Additionally, we demonstrate how to construct the partition associated with strong aggregation based on certain subgroups of the automorphism group. This approach can efficiently reduce the size of the state space which could be as large as the exponential of the number of nodes in the compatibility graph before the aggregation. Finally, we illustrate this methodology with examples based on simple compatibility graphs, such as rings and the group of rotations.

1 Introduction

Stochastic matching models have been introduced recently to describe the waiting time spent by items in a system until they leave with the arrival of a compatible item. Following [1], the Matching model is a triple (G, μ, Φ) such that

1. $G = (V, E)$ is an undirected graph whose vertices in V are classes of items and whose edges in E model the allowed matching of items. G is called the matching graph or the compatibility graph.
2. μ is a distribution of probability to represent the arrivals of items. Alternatively, one can consider a collection of Poisson processes for a continuous-time model,
3. Φ is a matching policy. When an arriving item matches several types of items already waiting, it gives a couple of items which is selected and leave the system.

The matching graph represents the classes of items (with a set of nodes V) and the compatibility among classes of items (with a set of edges E). Upon

---

This work is partially supported by FMJH Program PGMO.

arrival, an item is queued if no compatible items are present in the system. Matching occurs when two compatible items are present, and it is performed according to the matching discipline. Typical matching disciplines are First Come First Match (FCFM), Random, and Match the Longest Queue. Once they are matched, both items leave the matching system immediately.

Assuming independent Poisson arrivals of items, ordinary graphs (i.e., graphs without self-loops), and certain disciplines like FCFM, the general matching model is associated with a Markov chain in an infinite state space. Under these assumptions, a necessary and sufficient condition for stability and a product form solution (for FCFM) were proved in [2] and [1]. The general matching model proposed in [2] and [1] was considering a general undirected matching graph G. In this paper, it is assumed that the arrival of items occurs one at a time. It is important to avoid confusion with the Bipartite Matching Model (see for instance, [3] and references therein) where the matching graph is bipartite and two items of distinct classes arrive at the same time. Bipartite Matching Models were motivated by the analysis of the public housing [4] and the kidney exchanges [5,6]. The kidney exchange arises when a healthy person who wishes to donate a kidney is not compatible (blood types or tissue types) with the receiver. Two incompatible pairs (or maybe more) can form a cyclic exchange so that each patient can receive a kidney from a compatible donor.

The approach we present now was inspired by two results we obtained recently. First, we have established that there exists some performance paradox for FCFM matching models [7]. When new edges are added to the matching graph, the expectation of the total number of customers may decrease. We have provided some examples to show that it is not always true. Thus, adding flexibility to the system (more edges in the compatibility graph) does not always improve performance (the average number of waiting items increases). The main difficulty in proving such a result is that we cannot explicitly derive this average number for a general graph. However, we have proved that one can lump the Markov chains associated with some simple compatibility graphs and FCFM discipline. Second, a novel type of compatibility graph with FCFM discipline was proposed in [8] and independently in [9]. In these models, an item is compatible with another item of the same type. Thus, two items of the same type cannot exist within the system and they two items of the same type may disappear (it depends on the matching discipline). If all nodes have self-loops, the Markov chain associated with the model is finite and ergodic under some simple assumptions on the arrival probabilities. Additionally, we have proved in [9] that these chains have a product from steady-state distribution.

These two former results indicate that employing strong aggregation of Markov chains could be a valuable approach for studying models associated with disciplines other than FCFM. Specifically, we anticipate utilizing the structure of compatibility graphs to lump the chains effectively. Here, we present the first results of this novel approach. We restrict ourselves to compatibility graphs where all the nodes have self-loops and adopt the Random matching discipline. Therefore, the chain is finite but usually quite large, provoking the strong aggregation of the chain as a meaningful approach for deriving analytical results or numerically computing the steady-state distribution. Thus one of the motivation of this paper is to reduce the size of the state-space to allow the numerical computation of the steady-state distribution with a numerical tool like [10]. Note that the utilization of the automorphism group to reduce the complexity of a Markovian models has previously been proposed. For instance, in [11], the authors employed lumping techniques before solving a reduced number of ODEs in an SIS model. Another idea, proposed in [12], involved approximating the steady-state distribution through lumping based on local symmetries. To the best of our knowledge, the application of graph automorphism to precisely analyze matching models has not been studied so far. The technical content of the paper is as follows. We begin in the next section by introducing the model and defining relevant notation. We also prove that the Markov chain is not reversible, necessitating the development of new techniques. Section 3 delves into an illustrative example: a ring-shaped compatibility graph comprising n nodes. Here, we investigate the lumping process associated with a partition derived from the rotations of the states. We prove that, under certain assumptions on the arrival probabilities, the Markov chain exhibits strong aggregability. In Sect. 4, we extend this result to any compatibility graph with a non-trivial automorphism group, utilizing a partition based on this group and under some constraints on the item arrival probabilities. Finally, Sect. 5 is dedicated to a modular decomposition of compatibility graphs where the lumpability property holds. This decomposition relies once again on the automorphism group of the compatibility graph.

2   Notation and Assumptions

Let $G = (V, E)$ be the compatibility graph, where V denotes the set of nodes representing item types. We use n to denote the cardinality of V, implying the n distinct types of items. The set E comprises edges $(x, y) \in V \times V$, where (x, y) indicates compatibility between the item x and y, and they are instantaneously removed from the system if selected by the matching discipline. Without loss of generality, we assume that G is connected. We further assume that all the nodes in G have a self-loop (remember that this is a new assumption like in [9] and that this assumption was not considered in [1]). If x is a node from V, $\Gamma(x)$ denotes the set of neighbors of x in G. Following [1], states are often referred to as words and items are called letters to represent complex matching disciplines. By construction, all states in the matching system are independent sets. Indeed, if two letters are neighbors in the matching graph, they cannot both be waiting simultaneously in the matching system.

Definition 1 (Independent Set, IS) . *Let $G = (V, E)$ be a graph, an independent set of G is a set of nodes IS such that $IS \subset V$ and for all $i, j \in IS$, $(i, j) \notin E$.*

We consider matching discipline $RANDOM$ defined as follows:

Definition 2 (RANDOM discipline, RND) . *Assuming that a word and a letter are compatible (i.e., we can match the letter with at least one letter in the word), we delete one compatible letter from the word. This letter is chosen with a uniform distribution among the compatible letters.*

Example 1. *Consider a state with 1 letter a, 1 letter b and 1 letter c . Suppose that a letter t arrives and that t matches with both a and b but it does not match with c. Then, one deletes the letter a with probability 1/2 and the letter b with probability 1/2.*

Note 1. Let m be a state, and $m[i]$ is the quantity of item i in that state. Due to the loop, it can be 0 or 1.

Finally, we consider a discrete-time model based on a stationary probability distribution of arrivals α_i. At each time slot, one can observe an arrival of an item i with probability α_i or no arrival at all with probability α_V.

Under these assumptions, we can obtain a discrete-time Markov chain:

Property 1. *To represent the state of the system, we will use the number of letters. Specifically, we denote the state by vector $\boldsymbol{X} = (X_1, ..., X_n)$, where X_i represents the number of item i. $\boldsymbol{X} = (X_1, ..., X_n)$ is a discrete-time Markov chain. Due to the loops in G, the number of items in each type in the system is either 0 or 1, resulting in the states of the Markov chains being Independent Sets. Consequently, the chain is finite. Let E denote the empty state. It is clear that there exists a path from E to any state and from any state to E (it is sufficient to use the same sequence of arrivals in both cases due to the loop in the compatibility graph). Therefore, the chain is irreducible. Assuming $\alpha_V > 0$, the chain is aperiodic. See Fig1 as an example of a compatibility graph and the associated Markov chain. In conclusion, the Markov chain is ergodic. However, the number of independent sets (and hence the number of states) could be as large as 2^{n-1} (for instance, with a Star compatibility graph).*

Considering the symmetry of the graph of the Markov chain, a natural question is to check the reversibility of the Markov chain. Unfortunately, for some compatibility graphs and Random disciplines, the chain is not reversible, as shown in the following example.

Property 2. *Assume that the compatibility graph is a ring with 5 nodes. The Markov chain associated with this compatibility graph and Random matching discipline is not reversible.*

Proof: we verify that the local balance relations do not hold. Specifically, we examine these relations for states E, (1), (3) and (1, 3):

$$\alpha_1 \pi(E) = (\alpha_1 + \alpha_2 + \alpha_5)\pi(1) \text{ and } \alpha_3 \pi(E) = (\alpha_3 + \alpha_2 + \alpha_4)\pi(3), \quad (1)$$

and

$$\alpha_3 \pi(1) = (\alpha_2/2 + \alpha_3 + \alpha_4)\pi(1, 3) \text{ and } \alpha_1 \pi(3) = (\alpha_2/2 + \alpha_1 + \alpha_5)\pi(1, 3), \quad (2)$$

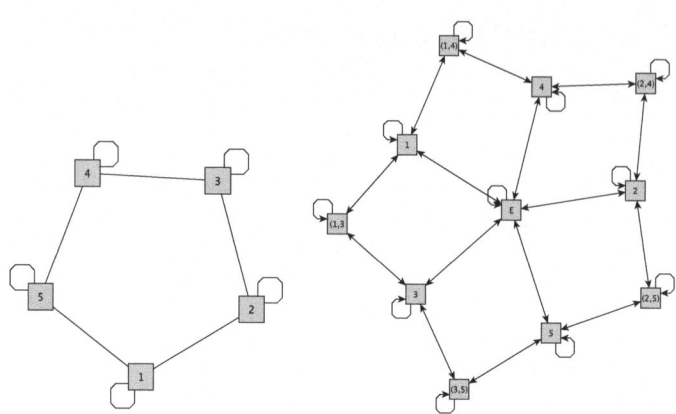

Fig. 1. Compatibility graph: Ring C_5 (left), associated Markov chain (right).

where the probability $\alpha_2/2$ comes from the Random discipline. Combining the first and the third relations to derive a relationship between π_E and $\pi(1,3)$ yields:

$$\pi_E = \pi(1,3)\frac{(\alpha_1 + \alpha_2 + \alpha_5)}{\alpha_1}\frac{(\alpha_2/2 + \alpha_3 + \alpha_4)}{\alpha_3}.$$

Performing the same operation with the second and fourth relations, we obtain:

$$\pi_E = \pi(1,3)\frac{(\alpha_3 + \alpha_2 + \alpha_4)}{\alpha_3}\frac{(\alpha_2/2 + \alpha_1 + \alpha_5)}{\alpha_3}.$$

The two numerators are not equal. This contradiction indicates that the chain is not reversible, as the local balance conditions do not hold.

It is important to note that while the chain may be reversible for certain graphs (such as the complete graph), the example provided demonstrated that this property does not hold for all of them. Consequently, we aim to establish some relations between the graph automorphisms and the lumping of the Markov chain associated with this compatibility graph. Before delving into this, let us first recall the definitions of lumpability and strong aggregation.

Definition 3 (Strong Aggregation). *Let W be a Markov chain on the set of states $\mathbb{W}$. Consider a partition $(B_1,..,B_k)$ of $\mathbb{S}$. We define a new process Y as follows:*

$$Y_n = m \leftrightarrow W_n \in B_m.$$

We aim to find conditions under which Y also forms a Markov chain. We denote Y as a strong aggregation of W for partition $(B_1,..,B_k)$.

The ordinary lumpability condition implies such a result

Definition 4 (ordinary Lumpability). *W is strongly lumpable for partition $(B_1,..,B_k)$ of its state space if, for all subset indices i and j, and for all states*

m_1 and m_2 in B_i, the following condition holds:

$$Pr(W_{n+1} \in B_j | W_n = m_1) = Pr(W_{n+1} \in B_j | W_n = m_2).$$

Here, B_i is denoted as macro-state i. Consequently, the blocks of the transition matrix associated with the macro-states of the partition exhibit a constant row sum.

Property 3. *Ordinary lumpability of the transition matrix is a sufficient condition for strong aggregation.*

The primary challenge lies in finding the appropriate partition and verifying the lumpability condition based on an abstract definition of the Markov chain (i.e., derived from the compatibility graph and the matching discipline). Before delving into the more general and formal result, we first present a simpler example to illustrate the techniques and the difficulties associated with this approach.

3   Ring Compatibility Graph C_n

We consider the graph G to be a ring with n nodes, each having a self-loop. We assume that the nodes are labeled from 0 to $n-1$ to simplify the definition of rotations.

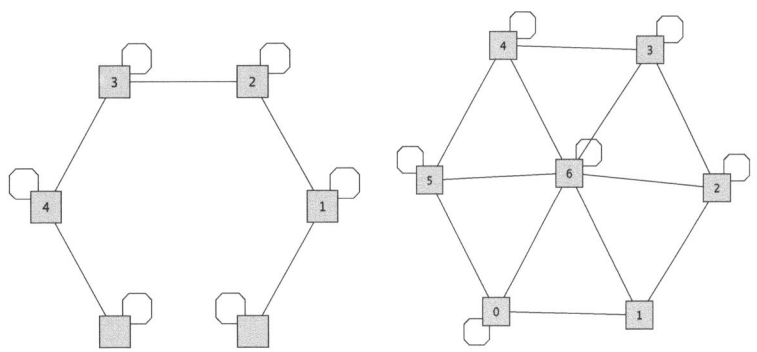

Fig. 2. Compatibility graph: (left Ring C_6), (right Wheel $W6$).

Note 2 (Operation $\oplus$. We denote $i \oplus j$ as $i + j$ mod n.

Definition 5. *The permutation σ_k on the set V of nodes of G is defined as $\sigma_k(i) = i \oplus k$. Clearly, $\sigma_k^{-1}(i) = i \oplus (n-k)$. Recall that the set of rotations forms a subgroup of the permutations.*

Note 3 Λ_k associated with σ_k is a permutation on the set of states S of the Markov chain, defined as $\Lambda_k(m) = (m_{\sigma_k^{-1}(0)}, ..., m_{\sigma_k^{-1}(n-1)})$.

Definition 6. *Let e_i be a state (represented as a vector) whose components are all equal to 0 except component i which is 1. Let m and p be some states, also considered as vectors, and the addition (denoted as "+") is performed component-wise.*

Property 4. *Let m and p be two words, we have for all k, $\Lambda_k(m+p) = \Lambda_k(m) + \Lambda_k(p)$. Additionally, for all k, we have $\Lambda_k(e_x) = e_{\sigma_k(x)}$.*

We consider the following partition $\mathcal{B}_m$, and we demonstrate the lumpability of the Markov chain for this partition when certain constraints on the arrival distributions are satisfied.

$$\mathcal{B}_m = \{p \in \mathcal{S} \mid \exists k, \ p = \Lambda_k(m)\}. \tag{3}$$

Clearly, the states in $\mathcal{B}_m$ are obtained by rotating the description of m.

The proof of ordinary lumpability requires that we can compute the transition probability between an arbitrary state within a macro-state and another macro-state. Hence, one must have an exhaustive list of states in every macro-state. However, we cannot explicitly construct $\mathcal{B}_m$ until we know the size of V. Therefore, we establish the following structural property, which is significantly stronger than ordinary lumpability.

First, we add the following condition on the arrival probability: we assume that $\alpha_i = \alpha = n$ for all i. This implies that $\alpha_x = \alpha_{\sigma_k(x)}$ for all k. Before proceeding with the proposition, it is essential to note that since graph G is a ring, each node has only three neighbors, including itself. Therefore, the number of possible transitions provoked by arrivals is relatively small: one inclusion, one deterministic deletion, or one deletion among two letters.

Lemma 1. *Let m be an arbitrary word and x a letter. Let p be a word in the same macro-state as m. Therefore, there exists k such that $\Lambda_k(m) = p$. Let $y = \sigma_k(x)$ (σ_k is the rotation associated with Λ_k, i.e., with the same k). Based on the type of transition, one can state two properties:*

- *If the action of x on m leads to a word r, then the action of y on p leads to a word s with the same probability. Furthermore, r and s are in the same macro-state.*
- *If the action of x on m leads to a word r uniformly drawn among $\{r_1, r_2\}$, then the action of y on p leads to a word s among $\{s_1, s_2\}$, with the same distribution of probability. Furthermore, r_1 and s_1 are in the same macro-state and a similar statement holds for r_2 and s_2.*

As we consider that σ is a rotation, we have $\sigma_k(x) = x \oplus k$. We have four cases to describe the action of letter x on word m and the proof must be done for all cases.

- Case 1: $x \notin m$ and $x \notin \Gamma(m)$. With probability 1, the letter x is added to the word m. The word becomes $m + e_x$.
- Case 2: $x \in m$. With probability 1, the letter x is deleted from the word m. The word becomes $m - e_x$.

– Case 3: $x \notin m$ and $|\Gamma(x) \cap m| = 1$. Let z be this letter. With probability 1, the letter z is deleted from the word m. The word becomes $m - e_z$.

– Case 4: $x \notin m$ and $|\Gamma(x) \cap m| = 2$. Let u and z be these letters. With probability $1/2$, letter z is deleted, and with probability $1/2$, letter u is deleted. Thus, the resulting word can be either $m - e_z$ or $m - e_u$, each with probability $1/2$.

Note that the first case does not depend on the graph G while the second one is related to the loops in G. The last two cases take into account that all the nodes of G have two neighbors. The last case is the only one related to the second property.

– Case 1: We begin with a more detailed description of the condition based on the properties of the graph and the state space of the Markov chain.
  - $x \notin m$ means that $m[x] = 0$
  - $x \notin \Gamma(m)$ means that $m[x \oplus 1] = 0$ and $m[x \oplus (n-1)] = 0$. Note that this is based on the graph edge set.

  We first need to prove the action of y on p also results in the arrival of a new letter in p. Thus we must prove that $p[y] = 0$, $p[y \oplus 1] = 0$ and $p[y \oplus (n-1)] = 0$. Given that $p = \Lambda_k(m)$, for all i, we have $p[i] = m[\sigma_k^{-1}(i)] = m[i \oplus (n-k)]$. Since $y = x \oplus k$, it means that $p[y] = m[x \oplus k \oplus (n-k)] = m[x]$. As $m[x] = 0$, we finally get $p[y] = 0$.

  Similarly, $p[y \oplus 1] = m[x \oplus k \oplus (n-k) \oplus 1] = m[x \oplus 1] = 0$. The proof is similar for $p[y \oplus (n-1)]$. Thus, the arrival of the letter y in state p also leads to the arrival of a letter in the word.

  Now, we need to check that these states belong to the same macro-state. By assumption, m and p belong to the macro-state $\mathcal{B}_m$. Therefore, there exists k such that $p = \Lambda_k(m)$. Let r be the state resulting from the arrival of x in m, and let s be the state resulting from the arrival of $\sigma_k(x)$ in $p = \Lambda_k(m)$. We need to prove that $s \in \mathcal{B}_r$. We have previously found that the effect of x on m is the addition of the letter x: $r = m + e_x$. Similarly, we have proved that $s = p + e_y$. Furthermore,

$$\Lambda_k(r) = \Lambda_k(m + e_x) = \Lambda_k(m) + e_{\sigma_k(x)} = p + e_y = s.$$

  Therefore, s belongs to $\mathcal{B}_r$, and case 1 is proved (see Fig. 3).

– Case 2: Again we can rewrite the condition $x \in m$ as $m[x] = 1$. We need to prove that the arrival of $y = \sigma_k(x)$ provokes a deletion in $p = \Lambda_k(m)$. For all i, we have $p[i] = m[\sigma_k^{-1}(i)] = m[i \oplus (n-k)]$. Similarly, $y = x \oplus k$ implies that $p[y] = m[x \oplus k \oplus (n-k)] = m[x] = 1$. Thus, $p[y] = 1$, and the arrival of y provokes a deletion in the word p. Let $r = m - e_x$ and $s = p - e_y$. We still need to prove that states r and s belong to the same macro-state. Again, we have $\Lambda_k(r) = s$ as

$$\Lambda_k(r) = \Lambda_k(m - e_x) = \Lambda_k(m) - e_{\sigma_k(x)} = p - e_y = s.$$

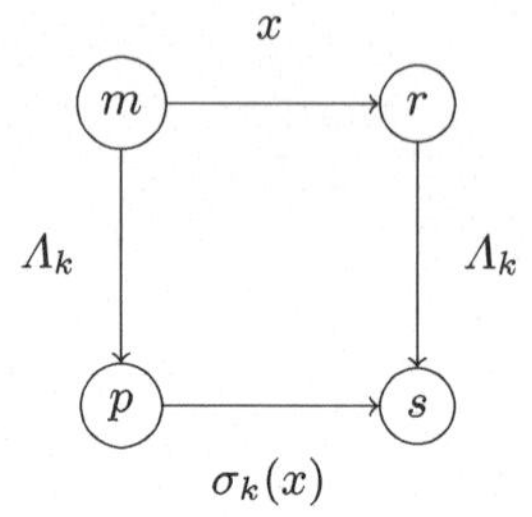

Fig. 3. Case 1 to 3, rotation.

Thus, case 2 is proved.

- Case 3: First, we rewrite the conditions as in the first two cases: $m[x] = 0$, $z = m \cap \Gamma(x)$, and the effect of letter x is a transition to state $m - e_z$. Without loss of generality, we assume that $z = x \oplus 1$ (the other case is $z = x \oplus n - 1$). Note that this takes into account the edge set of the graph.

Then, we have $m[x \oplus 1] = 1$, and the effect of letter x is a transition to $r = m - e_{x\oplus1}$ (i.e., component $x \oplus 1$ is now equal to 0). Now, consider $p = \Lambda_k(m)$ and the arrival of letter $y = \sigma_k(x)$ at state p. We have:

$$p[y \oplus 1] = p[x \oplus k \oplus 1] = m[x \oplus k \oplus 1 \oplus (n - k)] = m[x \oplus 1] = 1.$$

Similarly,
$$p[y] = p[x \oplus k] = m[x] = 0,$$

and
$$p[y \oplus (n - 1)] = p[x \oplus k \oplus (n - 1)] = m[x \oplus (n - 1)] = 0.$$

Therefore, the arrival of letter y provokes the transition from word p to word $s = p - e_{x \oplus k \oplus 1}$ with probability 1. Furthermore,

$$\Lambda_k(r) = \Lambda_k(m - e_{x\oplus1}) = \Lambda_k(m) - e_{x\oplus k\oplus1}.$$

Thus, we have $\Lambda_k(r) = s$. Therefore, r and s are in the same macro-state.

- Case 4: We have $m[x] = 0$, $\{u, z\} = m \cap \Gamma(x)$ and the effect of x is the deletion of letter u with probability $1/2$ (leading to state $r2$) or letter z with the same probability (leading to state $r1$). Without loss of generality, we assume that $z = x \oplus 1$ and $u = x \oplus (n - 1)$. Thus $m[x \oplus 1] = 1$ and $m[x \oplus (n - 1)] = 1$. Let $p = \Lambda_k(m)$ and $y = \sigma_k(x)$. Therefore, we have

$$p[y \oplus 1] = p[x \oplus k \oplus 1] = m[x \oplus k \oplus 1 \oplus (n - k)] = m[x \oplus 1] = 1,$$

and
$$p[y \oplus (n - 1)] = p[x \oplus k \oplus (n - 1)] = m[x \oplus (n - 1)] = 1,$$

and
$$p[y] = p[x \oplus k] = m[x] = 0.$$

Thus, the arrival of letter y provokes the deletion of letter $y \oplus 1$ with probability $1/2$, leading to state $s1$ or letter $y \oplus (n-1)$ with the same probability, leading to state $s2$ (see Fig. 4).

The same arguments as in Case 3 prove that $r1$ and $s1$ are in the same macro-state. Similarly, we have the same result for $r2$ and $s2$.

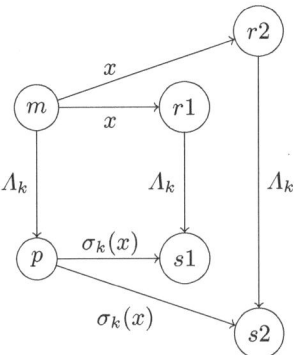

Fig. 4. Case 4, rotation.

Theorem 1 *Assume that for all* i, $\alpha_i = \alpha$. *The Markov chain* $(C_n, (\alpha_i)_{i=0..n-1}, RND)$ *is strongly aggregable for the partition defined in Eq.(3).*

Proof: To prove the ordinary lumpability, we consider two macro-states defined by Eq.(3): $\mathcal{B}_i$ and $\mathcal{B}_j$, two arbitrary states m and p in $\mathcal{B}_i$. Without loss of generality, we assume that there is a positive probability of transition from m to $\mathcal{B}_j$. By construction:

$$Pr(m, \mathcal{B}_j) = \sum_{l=0}^{n-1} \alpha_l \sum_{u \in \mathcal{B}_j} Pr(\text{arrival of letter } l \text{ provokes a transition from } m \text{ to } u).$$

But Lemma 1 implies that for all k,

$$\sum_{u \in \mathcal{B}_j} Pr(\text{arrival of letter } l \text{ provokes a transition from } m \text{ to } u) =$$

$$\sum_{u \in \mathcal{B}_j} Pr(\text{arrival of letter } \sigma_k(l) \text{ provokes a transition from } \Lambda_k(m) \text{ to } u).$$

As σ_k is a one-to-one mapping and $\alpha_l = \alpha$ for all letters l, we finally get that the lumpability condition holds.

4 An Arbitrary Graph with a Non-Trivial Automorphism Group

Let us first recall the definition of the automorphism group of a graph.

Definition 7 *Consider an arbitrary graph $G = (V, E)$. Let σ be a permutation on the set V. σ is a graph automorphism if, for all nodes u and v, $(u, v) \in E$ implies that $(\sigma(u), \sigma(v)) \in E$.*

We assume that G has a non-trivial automorphism group $\text{Aut}[G]$. We assume that V contains n nodes and that all nodes of G have a self-loop. We consider $\mathcal{A}$ a subgroup of $\text{Aut}[G]$. For instance, in the former section, the automorphism group of a ring with n nodes is the dihedral group, but we only considered the subgroup of rotations. The dihedral group also contains reflections. Let $\mathcal{S}$ be the state space of the Markov chain associated with compatibility graph G, arrival probability distribution α_i, and RANDOM discipline. Let σ be a permutation that is in $\mathcal{A}$ and let Λ_σ be a permutation of the states $m = (m_1, .., m_n)$ of $\mathcal{S}$, defined by

$$\Lambda_\sigma(m_1, .., m_n) = (m_{\sigma^{-1}(1)}, ..., m_{\sigma^{-1}(n)}).$$

Remember that as $\mathcal{A}$ is a subgroup of $\text{Aut}[G]$, the permutation σ^{-1} exists and is also in $\mathcal{A}$. We have the following property which slightly generalizes the previous propositions on rotations.

Property 5 *Let m and p be two words. For all automorphism σ in $\mathcal{A}$, we have:*

$$\Lambda_\sigma(m + p) = \Lambda_\sigma(m) + \Lambda_\sigma(p), \text{ and } \Lambda_\sigma(e_x) = e_{\sigma(x)}.$$

Now graph G is an arbitrary graph with a non-trivial automorphism group and self-loops on any node. We consider the following partition of the state space which is a straightforward generalization of the previous model.

$$\mathcal{B}_m = \{p \in \mathcal{S} \mid \exists \sigma \in \mathcal{A}, \ p = \Lambda_\sigma(m)\}. \tag{4}$$

Here, the states in $\mathcal{B}_m$ are obtained using any permutation of $\mathcal{A}$ on the description of m. Thus, changing the set $\mathcal{A}$ may lead to a larger number of states in a macro-state. However, it also increases the number of constraints on the arrival probabilities. Thus one can study the tradeoff between the size of $\mathcal{A}$, the constraints of α_i, and the number of macro-states. Again, as we are not able to explicitly build $\mathcal{B}_m$ to obtain the transition matrix, we propose to prove a harder property which implies ordinary lumpability.

Lemma 2. *Let m be an arbitrary word in $\mathcal{S}$, x a letter in V, σ a permutation in subgroup $\mathcal{A}$, and p a state in $\mathcal{B}_m$ associated with σ) (i.e., $\Lambda_\sigma(m) = p$. Let $y = \sigma(x)$. If the action of x on m leads to a word r, then the action of y on p leads to a word s with the same probability and r and s are in the same macro-state.*

To proceed with the proof, we need to consider the following three cases, each based on the action of letter x on word m:

- Case 1: $x \notin m$ and $\Gamma(x) \cap m = \emptyset$. In this scenario, as the discipline is greedy, the letter x is added to the word w with probability 1. Thus, the resulting word becomes $m + e_x$.
- Case 2: $x \in m$. Here, as the discipline is greedy, the letter x is deleted from the word w with probability 1. Consequently, the word becomes $m - e_x$.
- Case 3: $x \notin m$ and $W = \Gamma(x) \cap m$, where W is not empty.

Proof: it mimics the proof of Lemma 1.

- Let us first consider Case 1. The conditions are $m[x] = 0$ and for all $u \in \Gamma(x)$, we also have $m[u] = 0$. Let $p = \Lambda_\sigma(m)$ for an arbitrary σ in $\mathcal{A}$ and $y = \sigma(x)$. By construction, we have

$$p[y] = m[\sigma^{-1}(y)] = m[\sigma^{-1}(\sigma(x))] = m[x] = 0.$$

We aim to prove that $p[\sigma(u)] = 0$ for all u in $\Gamma(x)$. By definition:

$$p[\sigma(u)] = m[\sigma^{-1}(\sigma(u))] = m[u] = 0.$$

Recall that σ is an automorphism. Therefore, $(x, u) \in E$ if and only if $(\sigma(x), \sigma(u)) \in E$. Consequently, for all v in $\Gamma(\sigma(x))$, we have $p[v] = 0$. Thus, the arrival of letter y in p leads to state $p + e_y$ with probability 1. The crucial property we rely on here is the greediness of the matching discipline, ensuring that the arrivals of letter x or y cannot be discarded.
- Similarly to Case 1, Case 2 can be proved with a similar argument. In this case, we have $m[x] = 1$. Therefore,

$$p[y] = m[\sigma^{-1}(y)] = m[\sigma^{-1}(\sigma(x))] = m[x] = 1.$$

Thus, the arrival of letter y provokes the deletion of letter y already in state p. The resulting state is $p - e_y$ with probability 1. Once again, this case does not depend on the specific definition of the discipline. All that is required is that it is greedy.
- Now we have to consider the most challenging case: $x \notin m$ and $\Gamma(x) \cap m$ is not empty. There are three key points to address.
  First, by assumption, $m[x] = 0$. Thus, $p[y] = p[\sigma(x)] = m[\sigma^{-1}(\sigma(x))] = m[x] = 0$. Therefore, the letter y is not in the word p. Consequently, the arrival of y leads to either the insertion of y or the deletion of a neighbor of y in p.
  Second, let z be a letter of $\Gamma(x) \cap m$. Thus, $z \in \Gamma(x)$, or equivalently, (x, z) is an edge of E. Since σ is an automorphism, $(\sigma(x), \sigma(z))$ is also an edge of E and $\sigma(z) \in \Gamma(f(x))$.
  Furthermore, z is a letter of m, or equivalently, $m[z] = 1$. Thus, $p[\sigma(z)] = m[\sigma^{-1}(\sigma(z))] = m[z] = 1$.
  Combining both results, we have: $\sigma(z) \in \Gamma(\sigma(x)) \cap \Lambda_\sigma(m)$. Let $q = |\Gamma(x) \cap m|$. Since σ is a one-to-one mapping, we also have, for all $\sigma \in \mathcal{A}$

$$|\Gamma(\sigma(x)) \cap \Lambda_\sigma(m)| = q.$$

Finally, due to the RANDOM discipline, for the arrival of letter x in m or the arrival of letter $\sigma(x)$ in $\Lambda_\sigma(m)$, the transitions are described by the deletion of one letter (among q) with probability $1/q$.

Third, let $r_1, r_2, \ldots, r_q$ be the words obtained by arrival of letter x at state m (see Fig. 4) by removing letters $z_1, z_2, \ldots, z_q$. Consider the words s_i obtained by the arrival of letter $\sigma(x)$ at state $\Lambda_\sigma(m)$, we need to prove that we can order these words $s_1, s_2, \ldots, s_q$ such that $s_i = \Lambda_\sigma(r_i)$ (i.e. r_i and s_i are in the same macro-state). Indeed, we define r_i as $m - e_{z_i}$. Thus,

$$\Lambda_\sigma(r_i) = \Lambda_\sigma(m) + e_{\sigma(z_i)}.$$

Since z_i is a neighbor of x and σ is an automorphism, $\sigma(z_i)$ is a neighbor of $\sigma(x)$. Therefore, among the words $s_1, s_2, \ldots, s_q$, there exits one of them that is equal to $p + e_{\sigma(z_i)}$. Let us call this word s_i. Finally, we have $\Lambda_\sigma(r_i) = s_i$, proving that r_i and s_i are in the same macro-state. And the proof is complete.

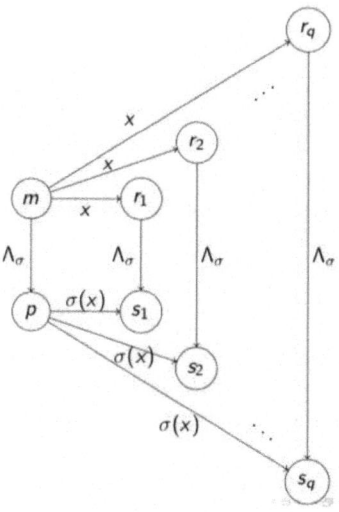

Fig. 5. Case 3, arbitrary graph.

Definition 8. *We define a set of constraints Z as a subset of $V \times V$ such that $(i, j) \in Z$ if and only if $\alpha_i = \alpha_j$.*

In the following, we assume some consistency between the set of constraints Z and the subgroup $\mathcal{A}$. Specifically, we assume that for all $\sigma \in \mathcal{A}$ and $i \in V$, we have $(i, \sigma(i)) \in Z$.

Theorem 2. *If the arrival probabilities α_i satisfy the constraints set Z, then the Markov chain $(G, (\alpha_i)_{i=1..n}, RND)$ is strongly aggregable for the partition defined in Eq.(4), based on the subgroup $\mathcal{A}$ of $\mathrm{Aut}[G]$:*

Proof: The first part of the proof mirrors that of Theorem 1. With the same notation, we consider

$$Pr(m, \mathcal{B}_j) = \sum_l \alpha_l \sum_{u \in \mathcal{B}_j} Pr(\text{arrival of letter } l \text{ provokes a transition from } m \text{ to } u).$$

As previously, Lemma 2 implies that the summations are equal:

$$\sum_{u \in \mathcal{B}_j} Pr(\text{arrival of letter } l \text{ provokes a transition from } m \text{ to } u) =$$

$$\sum_{u \in \mathcal{B}_j} Pr(\text{arrival of letter } \sigma(l) \text{ provokes a transition from } \Lambda_\sigma(m) \text{ to } u).$$

Clearly, if $\alpha_l = \alpha_{\sigma(l)}$ for all letters l, then $Pr(m, \mathcal{B}_j) = Pr(p, \mathcal{B}_j)$. The lumpability condition holds.

5   From G to $IN_1 \bowtie G$

We define the JOIN operation on compatibility graphs to construct larger graphs.

Definition 9. *The JOIN operation of two arbitrary graphs $G1 = (V1, E1)$ and $G2 = (V2, E2)$ yields graph $G = (V, E) = G1 \bowtie G2$:*

- *Nodes: $V = V1 \cup V2$,*
- *Edges: $E = E1 \cup E2 \cup \{(x, y), x \in V1, y \in V2\}$.*

Intuitively, we retain the nodes and edges of $G1$ and $G2$ and additionally include all edges between $V1$ and $V2$.

Lemma 3. *Let IN_1 be a compatibility graph with a single node x and a self-loop on this node. Consider an arbitrary graph G with a non-trivial automorphism group $\mathrm{Aut}[G]$. Then $\mathrm{Aut}[G \bowtie IN_1]$ is isomorphic to $\mathrm{Aut}[G]$.*

Proof: The edges of $G \bowtie IN_1$ consist of the edges of G along with the edges between node x and an arbitrary node in G. The automorphisms of $G \bowtie IN_1$ are the permutation that preserve node x and are automorphisms of G for the remaining part of the permutation. Therefore, $\mathrm{Aut}[G \bowtie IN_1]$ is isomorphic to $\mathrm{Aut}[G]$.

Let $\mathcal{A}$ be the subgroup of $\mathrm{Aut}[G]$ used for lumping. For an arbitrary permutation σ in $\mathcal{A}$, we construct a subgroup $\mathcal{B}$ of permutations whose cyclic decomposition is $(\sigma, 1)$.

Corollary 1 *If the Markov chain (G, α_i, RND) is strongly aggregable for the partition defined in Eq.(4), then the Markov chain $(G \bowtie IN_1, (\alpha_i, \alpha_x), RND)$ is strongly aggregable for the induced partition defined by $\mathcal{B}$ under the same conditions on α_i. No additional conditions are imposed on α_x.*

Example 2 *Consider the Wheel graph with 6 nodes on the periphery and node 6 at the center (say $W6$, in Fig. 2). Clearly, $W6 = C6 \bowtie IN_1$, and apply the previous results accordingly. Now, assume that $\alpha_1 = \alpha_3 = \alpha_5$, $\alpha_2 = \alpha_4 = \alpha_6$, $\alpha_7 > 0$, and $\alpha_V > 0$. Then, the Markov chain associated with this graph and these parameters are strongly aggregable for the partition associated with the set of permutations based on the following cycle decomposition $((1, 3, 5)(2, 4, 6), (7))$.*

6   Conclusions and Remarks

We now plan to explore compatibility graphs without self-loops, incorporating various matching disciplines. Deriving a straightforward generalization of the proofs presented here may not be feasible, given that the state space will be infinite. One of the primary advantages of compatibility graphs lies in the possible definition of much more complex matching disciplines. Indeed, many matching disciplines, such as Match the Longest, lack relevance when the number of items per class is limited to 0 or 1. We intend to extend some of the results obtained in this study to any intuitive greedy discipline, while also examining the tradeoff involved in computing subgroup $\mathcal{A}$, the constraints set Z, and reducing the size of the lumped state space to allow computation of the steady-state distribution.

References

1. Mairesse, J., Moyal, P.: Stability of the stochastic matching model. J. Appl. Probab. **53**(4), 1064–1077 (2018)
2. Moyal, P., Bušić, A., Mairesse, J.: A product form for the general stochastic matching model. J. Appl. Probab. **57**(2), 449–468 (2021)
3. Bušić, A., Gupta, V., Mairesse, J.: Stability of the bipartite matching model. Adv. Appl. Probab. **45**(2), 351–378 (2013)
4. Caldentey, R., Kaplan, E.H., Weiss, G.: FCFS infinite bipartite matching of servers and customers. Adv. Appl. Probab. **41**(3), 695–730 (2009)
5. United Network for Organ Sharing. https://unos.org/wp-content/uploads/unos/living_donation_kidneypaired.pdf
6. Unver, U.: Dynamic kidney exchange. Rev. Econ. Stud. **77**(1), 372–414 (2010)
7. Cadas, A., Doncel, J., Fourneau, J.-M., Busic, A.: Flexibility can hurt dynamic matching system performance. ACM SIGMETRICS Performance Evaluation Review, IFIP Performance evaluation (short paper) **49**(3), 37–42 (2021)
8. Begeot, J., Marcovici, I., Moyal, P., Rahme, Y.: A general stochastic matching model on multigraphs (2020). ArXiv preprint: https://arxiv.org/abs/2011.05169
9. Busic, A., Cadas, A., Doncel, J., Fourneau, J.-M.: Product form solution for the steady-state distribution of a Markov chain associated with a general matching model with self-loops. In: Gilly, K., Thomas, N. (eds.) Computer Performance Engineering: 18th European Workshop, EPEW 2022, Santa Pola, Spain, September 21–23, 2022, Proceedings, pp. 71–85. Springer International Publishing, Cham (2023). https://doi.org/10.1007/978-3-031-25049-1_5
10. Fourneau, J.M., Mahjoub, Y.A.E., Quessette, F., Vekris, D.: XBorne 2016: a brief introduction. In: Czachórski, T., Gelenbe, E., Grochla, K., Lent, R. (eds.) ISCIS 2016. CCIS, vol. 659, pp. 134–141. Springer, Cham (2016). https://doi.org/10.1007/978-3-319-47217-1_15
11. Simon, P.L., Taylor, M., Kiss, I.Z.: Exact epidemic models on graphs using graph-automorphism driven lumping. J. Math. Biol., pp. 479–508 (2011)
12. KhudaBukhsh, W.R., Auddy, A., Disser, Y., Koeppl, H.: Approximate lumpability for Markovian agent-based models using local symmetries. J. Appl. Probab. **56**(3), 647–671 (2019)

Optimal Allocation of Tasks to Networked Computing Facilities

Vincenzo Mancuso[1]([✉]) [iD], Paolo Castagno[2] [iD], Leonardo Badia[3]([✉]) [iD],
Matteo Sereno[2] [iD], and Marco Ajmone Marsan[1] [iD]

[1] IMDEA Networks Institute, Madrid, Spain
{vincenzo.mancuso,marco.ajmone}@imdea.org
[2] Department of Informatics, University of Turin, Turin, Italy
{paolo.castagno,matteo.sereno}@unito.it
[3] University of Padova, Padua, Italy
leonardo.badia@unipd.it

Abstract. Distributed allocation of computing tasks over network resources is meant to decrease the cost of centralized allocation. However, existing analytical models consider practically indistinguishable resources, e.g., located in the data center. With the rise of edge computing, it becomes important to account for the impact of diverse latency values imposed by edge/cloud data center locations. In this paper, we study the optimization of computing task allocation considering both the delays to reach edge/cloud data centers and the response times of servers. We explicitly evaluate the resulting performance under different scenarios. We show, through numerical analysis and real experiments, that differences in delays to reach data center locations cannot be neglected. We also study the price of anarchy of a distributed implementation of the computing task allocation and unveil important properties such as the price of anarchy being generally small, except when the system is overloaded, and its maximum can be computed with low complexity.

Keywords: Network servers · Optimization with network latency constraints · Next generation networking · Game Theory · Price of Anarchy

1 Introduction

Meeting latency requirements is fundamental to achieve the desired quality of service for real-time applications [23,28]. A paradigm often adopted in mobile Internet architectures to tackle this issue is that of edge computing, i.e., bringing computing resources closer to end users, rather than processing all data in the cloud [12]. However, as the mobile Internet becomes more pervasive, the management of distributed computing infrastructure is evolving towards an edge-cloud continuum, rather than a dichotomy between edge or cloud [3].

From the standpoint of an individual user, the problem is limited to the choice of the best (i.e., minimum latency) path [13]. When a global perspective is

© The Author(s), under exclusive license to Springer Nature Switzerland AG 2025
A. Devos et al. (Eds.): ASMTA 2024, LNCS 14826, pp. 33–50, 2025.
https://doi.org/10.1007/978-3-031-70753-7_3

adopted, establishing coordination among multiple users becomes of formidable complexity and is practically infeasible. The crux of the matter becomes whether distributed approaches to server selection in extremely variegate network architectures can still be efficient [1].

While this problem received attention in the past, we argue that the available frameworks are inadequate to represent the edge-cloud continuum. Most investigations consider a homogeneous latency model among the alternatives [10], so that the comparisons involve, e.g., fast vs slow servers but just with different parameters in the same formula. Instead, the alternatives in the mobile Internet are different not just in quantitative but also in qualitative terms.

One particular instance of this aspect is the fixed component of the latency, which comes for the most from the physical distance of the server in the edge-cloud continuum [15]. We will show how neglecting this aspect leads to suboptimal choices. To complicate things further, one can observe that, while service capacity and fixed latency can be possibly known to the user, the final performance depends on congestion at the chosen server, which is harder to estimate [25].

Following an algorithmic game theory pathway [19], we consider service choices made by a multitude of atomic non-cooperative agents interested in minimizing their own latency, and we derive the price of anarchy (PoA) [11]. Compared to results available in the literature, we take a general approach applicable to any functional relationship describing the latency, under very mild assumptions of positive first and second derivative. This results in a direct low-complexity implementation akin to water-filling algorithms [27]. Moreover, thanks to the generality of our approach, we can validate our quantitative results with real-world experiments.

Thus, we present multiple contributions: First, we evaluate the efficiency of distributed choices by network users in the edge-cloud continuum, under a general framework not found in the previous literature. We investigate the Nash equilibrium (NE) of distributed selections, which is found to be unique. Second, we formulate exact algorithms to find the optimal allocation point and the NE. Third, we compute the PoA as a function of network load. The worst-case PoA is proven to be the maximum among a finite number of cases. Eventually, we perform an extensive evaluation corroborated by experimental results.

2   Related Work

Server selection for a computing task typically represents and compares the servers as queues. The novelty of this paper is to include a fixed delay term to reach the selected server, after which the waiting and service time in the queue depend on the load as a general function with positive first and second derivatives. This resonates in the literature at many levels. To start, research in transportation optimization, led by Braess [2] and Pigou [17], has yielded significant findings that have applications in computer networks, routing, and server selection optimization. In-depth discussions can be found in [6] or [19].

A close alignment with our setup of server selection can be found in [1, 10, 26]. These works compare a selfish strategy, where users choose servers to minimize their own mean waiting time, against the social (i.e., global) optimization. In [16], the problem is addressed as a network routing problem, where each user aims to optimize its own performance. This leads to a non-cooperative game, with an emphasis on the conditions for an NE. Studies [1] and [10] analyze n exponential servers with FCFS discipline and develop a closed-form solution for the PoA. They consider service (i.e., queue response time) as the only parameter. If we introduce a fixed delay for each server, these PoA derivations become unfeasible.

Many more works exist in this area, since multiple authors worked on this and related issues. The highly influential work [18] led to scrutinizing the problem through diverse lenses, exploring numerous variants. For instance, an attempt to go beyond an exponential service time distribution is made in [26] for $M/G/1$ servers. An exact analysis of the PoA is presented only for a processor-sharing queuing discipline, where it is known that the average delay depends on the service time distribution only through its mean. Paper [26] highlights the mathematical difficulty of scenarios with FCFS $M/G/1$ servers, where it is inevitable to deal with choices depending on two parameters: mean and coefficient of variation of the service time distribution. [1] investigates a choice among $M/G/1$ servers focusing on the service rate and the coefficient of variation of the service time distribution. However, that analysis is valid only under the assumption that all servers share a common coefficient of variation. Another approach attempting, in various ways, to address server selection as characterized by two parameters is presented in [20], where heterogeneous costs and service times across different paths are studied. In there, the PoA is derived by applying simplifications or reducing the analysis to specific scenarios. A different approach that circumvents the use of two parameters in the service selection is presented in [21], with servers of type $M/G/1$ and $GI/GI/1$, but only under heavy traffic conditions that greatly simplifies the analysis.

Yet, we show that using multiple parameters is tractable. Indeed, we characterize each server with at least *two* independent parameters, i.e., delays in addition to service rates. Also, we generalize the delay function, as it can be defined by an arbitrary number of parameters, and admit, e.g., heterogeneous coefficients of variance, showing that obtaining exact solutions is still possible. In particular, we show that optimal solutions and NEs can be computed with exact algorithms in polynomial time.

Evaluations of the impact of the distance in cloud/edge scenarios generally pertain the problem of server placement [24]. More recently, with the evolution of multi-server architectures towards a horizontal edge-cloud continuum without hierarchy, the issue of choosing the servers from the individual user perspective has gained momentum [22]. However, most of the proposals advocate for heuristic low-complexity solutions or reinforcement learning strategies, which are found to be efficient. Our analysis shows that this does not happen by accident, but rather is grounded in the problem as characterized by useful structural properties [4] that we rigorously prove to hold even with the inclusion of the delay terms.

In conclusion, we present the first analysis that does not neglect the delay term to reach the servers—a practice that previous studies have regularly introduced, to make analytical expressions tractable. Our results solve a problem that has been open for many years, and avoid erroneous solutions to the server selection, computing instead the real optimal choices (in selfish or global terms).

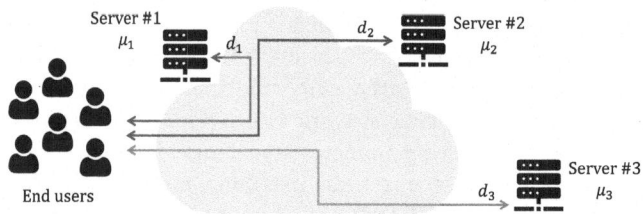

Fig. 1. Reference scenario: a group of users on a single network slice with computing resources accessible through the network in the edge-cloud continuum

3 Analysis with Fixed Path Delay

Assume that a group of users is connected to a network with n servers, as shown in Fig. 1. Although the servers are accessible at different distances, they belong to the same network slice, created and deployed over the edge-cloud continuum, as typical of modern 3GPP networks [5]. Servers are sorted in increasing order according to the average delay to serve a job, in turn determined by the sum of path latency and average service time, excluding queuing delays, that is,

$$\forall i, j \in \mathcal{S} = \{1, \cdots, n\}, \ i < j \implies d_i + 1/\mu_i \le d_j + 1/\mu_j, \tag{1}$$

where d_i is the two-way delay to reach server i and $1/\mu_i$ is the average service time at the same server. Offset values $d_i \ge 0$ make our study different from existing works, because they can make slow but close servers preferable to fast but far ones. Indeed, all previously published results only hold for the case $d_i = 0$ and some of them can be extended to the case $d_i = d_j$, $\forall i, j$.

In the following, $\mathbf{p}^\star = \{p_i^\star\}_{i=1,\cdots,n}$ denotes a probability vector that allocates load to servers so as to minimize the average latency, and $\mathbf{p}^\dagger = \{p_i^\dagger\}_{i=1,\cdots,n}$ denotes the probability vector at the NE, i.e., where the users selfishly split their load with a stochastic strategy, to minimize their own expected latency.

3.1 Optimal Load Allocation Problem

We aim to minimize the average system latency $U(\mathbf{p})$. If $\ell_i(x) > 0$ is the average latency of server i with capacity μ_i when it receives traffic with intensity $0 \le$

$x < \mu_i$, and Λ is the aggregate offered traffic, the average latency is

$$U(\mathbf{p}) = \sum_{i=1}^{n} p_i \, \ell_i(p_i \, \Lambda). \tag{2}$$

The above latency has to be minimized subject to the following constraints:

$$p_i \geq 0, \quad \forall i; \qquad p_i \Lambda \leq \mu_i, \quad \forall i; \qquad \sum_i p_i = 1. \tag{3}$$

Function $\ell_i(x)$ accounts for path latency (d_i), average service capacity (μ_i), and can also account for additional parameters (e.g., the variance of the service time). The minimum average latency must be the sum of distance plus one service interval, i.e., $\ell_i(0) = d_i + 1/\mu_i$. Besides, in systems subject to congestion, $\ell_i(x)$ monotonically increases with x. Also, $\ell_i(x)$ is a convex function, as commonly observed for the latency of a queueing system. Hence, $\ell_i(x)$ and its first and second derivatives are positive functions, which makes the problem strictly convex. Thus, the solution is unique.

Lagrangian—with multipliers α_i, β_i, and γ—and KKT necessary conditions for the optimal solution $\mathbf{p}^\star$, are as follows:

$$\mathcal{L} = \sum_i p_i \ell_i(p_i \, \Lambda) - \sum_i \alpha_i \, p_i + \sum_i \beta_i \, (p_i \, \Lambda - \mu_i) + \gamma \left(1 - \sum_i p_i\right);$$

$$\frac{\partial \mathcal{L}}{\partial p_i} = \ell_i(p_i \, \Lambda) + p_i \, \Lambda \ell'_i(p_i \, \Lambda) - \alpha_i + \Lambda \beta_i - \gamma = 0, \quad \forall i;$$

$$\alpha_i \, p_i = 0, \quad \beta_i \, (p_i \Lambda - \mu_i) = 0, \quad \forall i;$$

$$\gamma\left(1 - \sum_i p_i\right) = 0.$$

Notice that $\ell_i(x) + p_i \, \Lambda \ell'_i(x)$ is the derivative of the weighted latency of the i-th server, with weight p_i. Multipliers α_i and β_i must be non-negative, while γ can take any real value (because it is associated to an equality constraint).

To identify which servers will receive traffic, consider the KKT conditions at $\mathbf{p}^\star$, which is a solution of the formulated problem. There are two cases:

Case $p_j^\star = 0$. If the optimal solution consists in assigning zero load to server j, then the KTT conditions imply that

$$\alpha_j \geq 0, \quad \beta_j = 0, \quad \ell_j(0) - \alpha_j - \gamma = 0, \quad \Rightarrow \ell_j(0) \geq \gamma, \tag{4}$$

i.e., servers that do not need to be active are those for which the latency computed at empty queue is not less than γ. However, with no queueing, a job is served in $\ell_j(0) = d_j + 1/\mu_j$, for any function ℓ_j. Therefore, if server j is not active, then server $i > j$, for which $d_i + 1/\mu_i \geq d_j + 1/\mu_j$, is also inactive. Being the set of servers ordered, there exists an integer $j^\star \leq n$ such that all servers from 1 to $j^\star$ receive some traffic, while the remaining ones are inactive.

Finding γ and $j^\star$ is part of the optimization. Moreover, their values are related, as it will emerge from the analysis of KKT conditions for $p_j^\star > 0$.

Case $p_j^\star > 0$. For all nodes $j = 1, \cdots, j^\star$, which receive non-zero load in the optimal configuration, the following conditions must hold:

$$\alpha_j = \beta_j = 0, \qquad \gamma = \ell_j(p_j^\star \Lambda) + p_j^\star \Lambda \ell_j'(p_j^\star \Lambda). \tag{5}$$

Thus, in a sense, we are following a water-filling approach on the derivative of the weighted latency. Moreover, term γ is larger than the largest latency of any server, and thus we will refer to it as to the *augmented latency*. The latter must also be positive and, since $p_j^\star > 0$, must also be greater than any $\ell_j(0)$ for any active server, which leads to

$$\gamma > d_{j^\star} + 1/\mu_{j^\star}. \tag{6}$$

These results are coherent with (4), which defines when a server receives no load. They also prove γ to be a monotonically increasing—hence invertible—function $h_j(x) = \ell_j(x) + x\ell_j'(x)$, which has to be computed at $x = p_j^\star \Lambda$ at the optimum. If h_j^{-1} indicates the inverse function, we have

$$p_j^\star = h_j^{-1}(\gamma)/\Lambda. \tag{7}$$

Considering that the latency functions ℓ_j and the corresponding derivatives are defined only for offered traffic comprised between 0 and μ_j, the above function $h_j^{-1}(\gamma)$ can only take values in the interval $[1, \mu_j]$ (or $[1, \mu_j)$, if the latency function diverges at μ_j). Hence, h_j^{-1}/Λ cannot be larger than 1. Indeed, the sum of all non-zero probabilities must be 1, so:

$$1 = \sum_{j=1}^{j^\star} p_j^\star = \frac{1}{\Lambda} \sum_{j=1}^{j^\star} h_j^{-1}(\gamma). \tag{8}$$

Since the sum of increasing functions is also increasing, the RHS of (8) is invertible. In addition, the RHS must be between 0 and $\sum_{j=1}^{j^\star} \mu_j/\Lambda \geq 1$, so that all the offered traffic can be served. Thus, there exists a unique γ satisfying (8).

Because of conditions (4) computed on $j^\star+1$ and (6) for $j^\star$, γ must be found in a specific interval, i.e., $\gamma \in \left(d_{j^\star} + \frac{1}{\mu_{j^\star}}, d_{j^\star+1} + \frac{1}{\mu_{j^\star+1}} \right]$, where $d_{j^\star+1} + \frac{1}{\mu_{j^\star+1}}$ has to be taken as infinite if $j^\star = n$.

Determining $j^\star$ beforehand is key to compute the problem solution efficiently. Indeed, once $j^\star$ is known, by inverting normalization (8), one can compute γ, which in turn can be used in (7) to compute the optimal probabilities $p_j^\star$.

To find $j^\star$, we need to consider that probabilities $p_j^\star$ depend on Λ, although so far we have treated Λ as a constant. Notice that γ must be monotonically increasing in Λ, because so are all ℓ_j and the associated derivatives that define

γ through (5). Thus, as traffic Λ increases, the augmented latency γ increases too, so that servers are progressively activated following the order of (1). Thus, when server j is activated at $\Lambda_j^{(\mathrm{Opt})}$, load $p_j^\star$ is still 0 and, according to (5), $\gamma = \ell_j(0) = d_j + \frac{1}{\mu_j}$. Hence, at $\Lambda_j^{(\mathrm{Opt})}$ we have:

$$p_i^\star \Lambda_j^{(\mathrm{Opt})} = h_i^{-1}\left(d_j + \frac{1}{\mu_j}\right), \quad \forall i \leq j, \tag{9}$$

and by summing over servers with non-zero probability $p_i^\star$, i.e., from 1 to $j-1$, we obtain the traffic threshold:

$$\Lambda_j^{(\mathrm{Opt})} = \sum_{i=1}^{j-1} h_i^{-1}\left(d_j + \frac{1}{\mu_j}\right). \tag{10}$$

A comparison between the activation thresholds $\Lambda_j^{(\mathrm{Opt})}$ and the offered traffic Λ thus reveals the value of $j^\star$. Of course, $\Lambda_1^{(\mathrm{Opt})} = 0$ since at least the first server has to be active as soon as non-zero traffic is offered to the network.

This optimization requires the inversion of a few monotonic functions, which admits closed form only in specific cases; in general, it can be done numerically with lightweight algorithms such as the dichotomous search. The following MIND-IT(Opt) (Minimum Delay Independent of Traffic and Service - Optimum) algorithm finds probability vector $\mathbf{p}^\star$ minimizing the average system latency.

MIND-IT(Opt)

Input: Sorted servers $j \in \{1, \cdots, n\}$, Λ.
Step 1: Compute activation thresholds $\Lambda_j^{(\mathrm{Opt})}$ with (10).
Step 2: Compute $j^\star$ by comparing Λ to the thresholds.
Step 3: Set $p_j^\star = 0, \forall j > j^\star$ and compute γ by inverting (8).
Step 4: Compute $p_j^\star$, $\forall j \leq j^\star$, with (7).
Output: $p_j^\star, j \in \{1, \cdots, n\}$.

Theorem 1. *The MIND-IT(Opt) algorithm is exact and polynomial.*

Proof. The optimality of feasible $\mathbf{p}^\star$ follows from it satisfying all KKT conditions. Evaluating the activation thresholds requires computing a number of terms quadratic in n and inverting (8). The latter consists of finding the zero of a monotonically increasing function with up to $n+1$ invertible terms, which can be done, e.g., with a dichotomous search on each one, with complexity $\mathcal{O}(n \log r)$, where r is the target numerical resolution. The complexity of computing each of the n probabilities, with (7), is that of one inversion, i.e., $\mathcal{O}(\log r)$. Therefore, the complexity of MIND-It(Opt) is $\mathcal{O}(n^2 + n \log r)$, polynomial for any r.

3.2   The Nash Equilibrium

We can consider a distributed version for the minimization of (2), in which each user sending traffic minimizes her latency with a probabilistic strategy $\mathbf{p_u^\dagger}$.

With sorted servers, a user sends traffic to j only if the observed average latency is above the minimum possible latency at that server, $\ell_j(0)$. An NE exists and is unique due to $U(\mathbf{p})$ being strictly convex [1]. Such a NE can be approached through subsequent approximations, in a water-filling form [7]. At the NE, all users choose a strategy $\mathbf{p_u^\dagger} = \mathbf{p^\dagger}$, which is also the same for all users due to symmetry, where the servers that receive traffic experience the same average latency τ, so that no user has an incentive to deviate from $\mathbf{p^\dagger}$:

$$\ell_j(p_j^\dagger \Lambda) = \tau, \qquad p_j^\dagger > 0, \qquad \forall j \leq j^\dagger, \tag{11}$$

where $j^\dagger$ is the number of used servers ($p_j^\dagger = 0$ for other servers $j > j^\dagger$).

As Λ increases, τ has to increase as well because, to maintain the same latency at all active servers, the incremental arrival rate has to be distributed over all of them and no server can receive less traffic than before the increase. Notice that, for a server j with $p_j^\dagger > 0$, the latency's lower bound is

$$\tau > \ell_j(0) = d_j + 1/\mu_j, \qquad \forall j \leq j^\dagger. \tag{12}$$

Expressing probabilities versus the average latency τ and normalizing, we obtain:

$$p_j^\dagger = \ell_j^{-1}(\tau)/\Lambda, \quad \forall j \leq j^\dagger; \tag{13}$$

$$1 = \sum_{j=1}^{j^\dagger} \ell_j^{-1}(\tau)/\Lambda, \tag{14}$$

where ℓ_j^{-1} indicates the inverse function of ℓ_j.

Inverting the above expression yields the value of $\tau > 0$, which must be unique since the RHS of (14) is monotonic increasing and upper-bounded by $(1/\Lambda)\sum_{j=1}^{j^\dagger} \mu_j \geq 1$, as the traffic offered to each server cannot exceed the capacity (i.e., $\ell_j^{-1}(\tau) \leq \mu_j$) and Λ cannot exceed the aggregate capacity of active servers. Notice that τ must be $d_{j^\dagger} + 1/\mu_{j^\dagger}$ when server $j^\dagger$ gets switched on, and as soon as $p_{j^\dagger}^\dagger$ becomes larger than zero, τ must be comprised in the interval

$$\tau \in \left(d_{j^\dagger} + 1/\mu_{j^\dagger}, d_{j^\dagger+1} + 1/\mu_{j^\dagger+1}\right), \quad \text{at } p_{j^\dagger}^\dagger > 0 \text{ and } p_j^\dagger = 0 \; \forall j > j^\dagger. \tag{15}$$

This implies that the average latency τ has to increase when a new server is activated because of an increase of Λ. We conclude that τ increases with Λ and servers are progressively activated following the sorting order (1), as Λ increases.

Since at activation of j the value of $p_j^\dagger$ is 0 and therefore $\tau = d_j + 1/\mu_j$, and the sum of non-zero probabilities must be equal to 1, the threshold $\Lambda_j^{(\text{NE})}$ can be computed similarly to the threshold in the optimal case (of course, $\Lambda_1^{(\text{NE})} = 0$):

$$\Lambda_j^{(\text{NE})} = \sum_{i=1}^{j-1} \ell_j^{-1} \left(d_i + 1/\mu_i\right).$$ (16)

In the NE calculation, function ℓ_j plays the role that h_j plays in the calculation of the optimum. Since both functions are positive and increasing, the algorithm for finding the NE is very similar, as shown in the following MIND-IT(NE) (Minimum Delay Independent of Traffic and Service – at the NE) algorithm.

MIND-IT(NE)

Input: Sorted servers $j \in \{1, \cdots, n\}$, Λ.
Step 1: Compute thresholds $\Lambda_j^{(\text{NE})}$ with (16).
Step 2: Compute $j^\dagger$ by comparing Λ to the thresholds.
Step 3: Set $p_j^\dagger = 0, \forall j > j^\dagger$ and compute τ inverting (14).
Step 4: Compute $p_j^\dagger$, $\forall j \le j^\star$, with (13).
Output: $p_j^\dagger, j \in \{1, \cdots, n\}$.

Theorem 2. *The MIND-IT(NE) calculation algorithm is polynomial.*

Proof. The proof proceeds as for Theorem 1 and finds complexity $\mathcal{O}(n^2 + n \log r)$ for any inversion precision r.

3.3 Properties

Here we present a few interesting properties, which will help comparing NE and optimal working points as the offered traffic changes. Proofs are omitted due to lack of space, but can be found in [14] jointly with more properties.

Lemma 1. *The augmented latency γ is an upper bound for the optimal latency. The distance of the bound is proportional to the derivative of optimal latency, which is a continuous function of traffic Λ:*

$$\frac{d}{d\Lambda} U(\mathbf{p}^\star) = \frac{\gamma - U(\mathbf{p}^\star)}{\Lambda} = \sum_{j=1}^{j^\star} \left(p_j^\star\right)^2 \ell_j'(p_j^\star \Lambda) > 0.$$ (17)

Beside being an upper bound for the optimal latency, it is useful to see that γ is also larger than the latency at the NE, as expressed in the following lemma.

Lemma 2. $\Lambda > 0 \implies \gamma > \tau$.

Lemma 3. $\Lambda_j^{(\text{NEP})} \ge \Lambda_j^{(\text{Opt})}, \forall j \in \{1, \cdots, n\}$.

Lemma 4. $\tau' \ge 0$ and $\tau'' \ge 0$ at any traffic $\Lambda \ge 0$.

3.4   Price of Anarchy

The PoA in the studied system is the ratio between average latency at the NE and average latency at the optimum. It is a function of the offered traffic—denoted as $\eta(\Lambda)$—and can only assume values greater than or equal to 1:

$$\eta(\Lambda) = \frac{U(\mathbf{p}^\dagger)}{U(\mathbf{p}^\star)} = \frac{\tau}{U(\mathbf{p}^\star)} \geq 1. \tag{18}$$

Conjecture 1. The PoA curve vs Λ is piece-wise convex.

The conjecture is motivated as follows. The PoA is the ratio of continuous functions of Λ, with the additional property that the denominator has also continuous and positive derivative because the condition used to compute γ at the optimum preserves the continuity of the derivative (cf. also Lemma 1). The numerator is non-decreasing and convex (cf. Lemma 4), but its derivative can be discontinuous at the activation points of servers at the NE, which occur after the activation of servers at the optimum (cf. Lemma 3). Therefore, consider the adjacent segments $[\Lambda_j^{(NE)}, \Lambda_{j+1}^{(NE)}]$. At $\Lambda_j^{(NE)}$, server j is activated and the NE assigns probabilities $\mathbf{p}^\dagger$ so that the curve of τ vs Λ be continuous. Immediately before and after the activation of server j the same latency is observed with a different number of servers, so that the last activated server j absorbs some load and the rest of servers observe a slowdown in the rate of increase of their latency. This corresponds to a sudden decrease of the derivative of τ. This means that while τ vs Λ experiences a drop in its growth rate at any point at which a new server is incorporated in the NE, the latency at the optimum does not observe such a discontinuity in the derivative. The result is that, for arrival rates slightly larger than $\Lambda_j^{(NE)}$, the PoA can decrease. However, the growth rate of τ must quickly increase again and faster than the growth rate of the latency at the optimum because the NE does not allow any server, not even the faster, to experience less latency than the slower. Therefore, the growth rate of τ is that of the slowest active server, while at the optimum this cannot occur by construction. Since the numerator of the PoA increases faster than the denominator, the PoA curve between two consecutive server activations at the NE must be convex. This behavior holds for all feasible load segments, including from $\Lambda_n^{(NE)}$ to $\sum_{j=1}^{n} \mu_j$.

If the above conjecture holds, as confirmed by our experiments, then searching for the worst-case PoA becomes simple.

Result 1. *The maximum of the PoA vs Λ occurs at $\Lambda > 0$, at a point of activation of a server at the NE or at $\rho = 1$.*

Result 1 tells that finding the maximum PoA can be done by evaluating a finite set of points, each of which requires to run in polynomial time the exact algorithms MIND-IT(NE) and MIND-IT(Opt). Hence, the cost of evaluating the worst case behavior of the distributed approach is polynomial and comparable to the complexity of the exact algorithms for the optimum and NE.

4   Special Case: M/M/1 Queues

With **M/M/1 queues**, ℓ_i and h_i can be inverted in closed form as

$$\ell_i(x) = d_i + \frac{1}{\mu_i - x}; \qquad h_i(x) = d_i + \frac{\mu_i}{(\mu_i - x)^2}. \tag{19}$$

General case expressions for optimization simplify into:

$$\gamma = d_j + \frac{\mu_j}{(\mu_j - p_j^\star \Lambda)^2}, \quad p_j^\star > 0, \quad \forall j \le j^\star; \tag{20}$$

$$p_j^\star = \frac{1}{\Lambda}\left(\mu_j - \sqrt{\frac{\mu_j}{\gamma - d_j}}\right), \quad \forall j \le j^\star; \tag{21}$$

$$\frac{1}{\Lambda}\sum_{j=1}^{j^\star}\left(\mu_j - \sqrt{\frac{\mu_j}{\gamma - d_j}}\right) = 1. \tag{22}$$

The complexity of inverting (22) is $\mathcal{O}(\log r)$ instead of $\mathcal{O}(n \log r)$ observed in the general case, because the inversion can be done over the sum directly. However, the overall complexity of the exact optimization remains $\mathcal{O}(n^2 + n \log r)$.
The expressions needed to study the NE become:

$$\tau = d_j + \frac{1}{\mu_j - p_j^\dagger \Lambda}, \quad p_j^\dagger > 0, \quad \forall j \le j^\dagger; \tag{23}$$

$$p_j^\dagger = \frac{1}{\Lambda}\left(\mu_j - \frac{1}{\tau - d_j}\right), \quad \forall j \le j^\dagger; \tag{24}$$

$$\frac{1}{\Lambda}\sum_{j=1}^{j^\dagger}\left(\mu_j - \frac{1}{\tau - d_j}\right) = 1. \tag{25}$$

In the very special case of **M/M/1 queues with equal delays** $d_j = d^\star$, closed forms can be found from the above formulas, which generalizes the conclusions of [10] for the optimization and NE of systems with M/M/1 queues with $d^\star = 0$. In particular, if $d_j = d^\star$ for all active servers, then we obtain:

$$p_i^\star = \frac{1}{\Lambda}\left(\mu_i - \sqrt{\mu_i}\frac{\sum_{j=1}^{j^\star}\mu_j - \Lambda}{\sum_{j=1}^{j^\star}\sqrt{\mu_j}}\right), \tag{26}$$

$$U(\mathbf{p}^\star) = d^\star + \frac{1}{\Lambda}\left(\frac{\left(\sum_{j=1}^{j^\star}\sqrt{\mu_j}\right)^2}{\sum_{j=1}^{j^\star}\mu_j - \Lambda} - j^\star\right). \tag{27}$$

The price of anarchy with M/M/1 and equal fixed delays is

$$\eta = \left(d^\star + \frac{j^\dagger}{\sum_{j=1}^{j^\dagger} \mu_j - \Lambda} \right) \Big/ \left(d^\star + \frac{1}{\Lambda} \left(\frac{\left(\sum_{j=1}^{j^\star} \sqrt{\mu_j} \right)^2}{\sum_{j=1}^{j^\star} \mu_j - \Lambda} - j^\star \right) \right). \qquad (28)$$

We conclude that, with M/M/1 queues and homogeneous delays $d_j = d^\star$, all quantities of interest can be expressed in closed form. However, this is in general not true when fixed latency terms d_j are not homogeneous. Nonetheless, the PoA at the border of the stability region of the system can be always computed in closed form, as shown in the following theorem.

Theorem 3. *With M/M/1 queues,*

$$\lim_{\Lambda \to \sum_{j=1}^{n} \mu_j} \eta(\Lambda) = n \sum_{j=1}^{n} \mu_j \Big/ \left(\sum_{j=1}^{n} \sqrt{\mu_j} \right)^2. \qquad (29)$$

Proof (Sketch). The average latency at the NE tends to diverge and can be computed in closed form by neglecting the fixed delay. By comparing $p_j^\star$ vs γ in near-saturation conditions, with $\gamma \gg d_j$, and summing all probabilities, we get the asymptotic expression for γ, hence obtain $p_j^\star$ and the average optimal latency.

5   Performance Evaluation

Experimental Platform — We set up a distributed measurement apparatus to obtain a ground truth and compare it with the predictions of our model. The apparatus enables the configuration of a group of servers (3 in our evaluations) with varying service capacities and diverse locations. This tool is coded in Golang [8], utilizing microservices for deployment, and employs QUIC as transport layer protocol [9]. The tool specifies three entities: clients, servers, and routing nodes. The server instantiates a configurable web processor that exposes one or multiple services with distinct computational demands. Client and server support a fine-tuning of applications and traffic shape characteristics. The routing element manages the traffic generated by other entities, directing it to its destination.

The tool collects and stores networking and routing events, including packet arrivals and departures to and from various elements, real-time monitoring of memory occupancy, and the count of available threads dedicated to handling incoming traffic. It also enables deploying application-specific measurements.

The experimental setting uses 3 servers deployed across Europe, each with deterministic service times, practically no bounds on available memory, and one working thread. The application requires clients to issue requests of 100 bytes to any available server; the server replies to each request by sending it back.

In our setting, several co-located clients issue packets to the three servers; the first two elements generate a traffic load equal to 5% of the overall system capacity—and this is the traffic being measured—while the last element provides background traffic. We measure the round trip time between each client and the three servers; then, we use it to compute the allocation of traffic to servers, and we feed this configuration to the routing element. The NE is found directly under the assumption of rational players, computing it beforehand.

Results and Validation — We used a few significant scenarios to validate our analytical model and showcase the results derived in this paper. Due to space limitations, we only reports results for model validation and with simple queueing discipline, stressing the importance of taking into account fixed delays in the presence of servers deployed in the edge-cloud continuum.

Scenario 0—Validation. We start by validating our model via experiments in a real network context. We observed the two-way delay between users and servers with ping packets, and used the average RTT (round trip time) to run our analytical model. The characteristics of the available servers, in order of activation, are $d = [20, 34, 43.5]$ ms, and $\mu = [4.66, 5.00, 10.20]$ services/s.

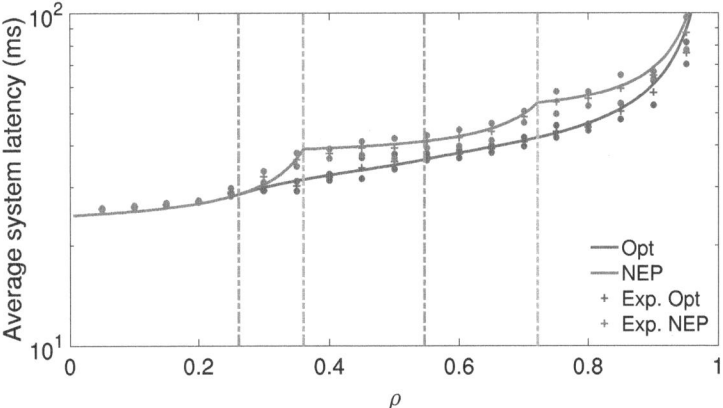

Fig. 2. Validation in Scenario 0

Figure 2 reports results generated by model and experiments (averages are reported as crosses, while 90% confidence intervals are delimited by dots). The figure also shows the activation thresholds of the servers, with vertical dotted lines (using the same color as the corresponding latency curves). Thresholds are computed analytically, based on the observed average two-way delay.

Analytical results show that differences between the optimum and the NE are generally small and can only be experienced when at least two servers are active. Experimental results match the analysis, which tells that considering a constant two-way delay instead of a stochastic model yields an affordable simplification.

Scenario 1—M/M/1 Edge & Cloud. We consider three M/M/1 servers with different capacity. One of the three servers is much farther apart from users than the other two servers, but it is faster. This represents a case in which two servers are within the edge area of the network and one is in the cloud. The characteristics of the available servers, in order of activation, are $d = [40, 30, 150]$ ms, and $\mu = [15, 9, 20]$ services/s. Latency and PoA as functions of the offered load in this scenario are shown in Fig. 3. As the second server gets activated at the optimum, the latency at the NE starts increasing faster than in the optimized system, and the PoA becomes larger than 1. However, the activation of the second server at the NE causes a temporary decrease in the PoA, after which it goes up quickly. The PoA curve is visibly piece-wise convex and the peak of the PoA is reached at the activation of the last server at the NE. The corresponding value is neither negligible nor very large (below 1.15). It is interesting that the PoA can be quite different at different loads, and a distributed implementation of the job allocation could work well at quite high loads.

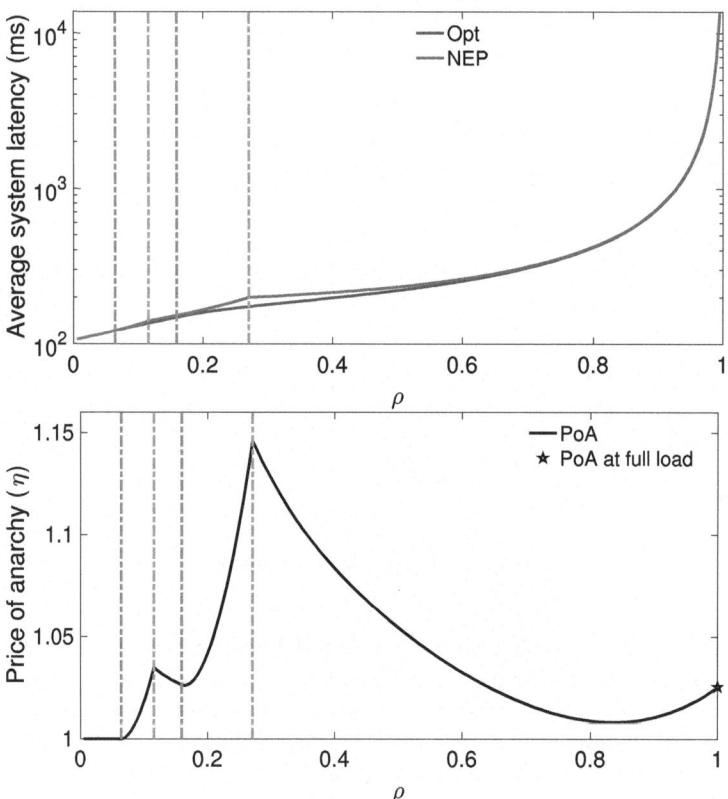

Fig. 3. Scenario 1. Top: latency, bottom PoA, vs offered load. The marked point indicates the PoA value computed with (29)

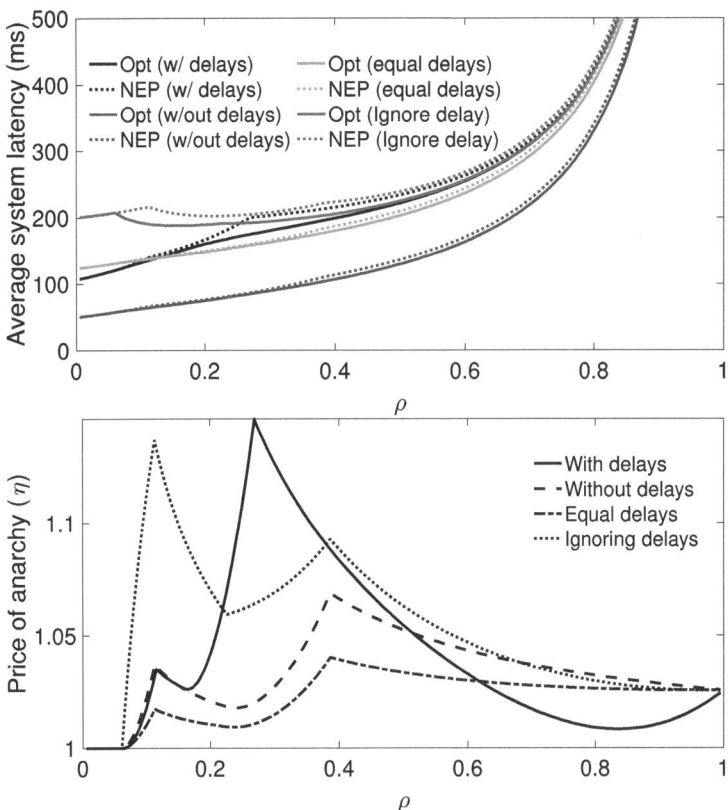

Fig. 4. Scenario 2. Left: latency, right PoA, vs offered load with variants on the evaluation and consideration of fixed delays

Scenario 2—The Importance of Accounting for Fixed Delays. Figure 4 reports results for a network configuration like the one of Scenario 1, although the optimization and the NE are computed under different assumptions and circumstances.

We first compare results discussed for Scenario 1 with the case in which fixed delays d_j are actually set to zero ("Without delays", in the figures). In this case, results are sensibly better than in the originally considered setup ("With delays"), at least in terms of latency, and only until the load becomes high. This means that fixed delays play an important role and neglecting them is not possible before the average sojourn time into servers becomes predominant.

Secondly, we consider a modified Scenario 1 in which fixed delays are equal for all servers (labeled as "Equal delays" in the figures). Such a delay is set to the average value observed in the original Scenario 1 (i.e., we set here $d_j = 73.3$ ms for all servers). Latency curves in this case are smoother than for Scenario 1, although the difference with Scenario 1 and with the case without fixed delays vanishes as the load approaches 1. Hence, the heterogeneity of fixed delays plays a non-negligible role as well, at least for low-medium loads.

We eventually consider the case in which the existence of fixed delays is neglected. This case is labeled as "Ignoring delays" in the figures. We remark that fixed delays are present in this case, and therefore latency and PoA values reported in the figures do account for their presence, although neither optimization nor NE calculation considered them (as if they were set to zero although they are not). Notice that this approach reflects exactly what analytical state-of-the-art solutions could do in similar cases. Interestingly, obtained curves shows large errors with respect to the correct traffic allocation optimization or NE calculation, which remarks how accounting for the presence of fixed delays makes a huge difference.

Note that the dotted curve in the figure, obtained by ignoring the presence of fixed delays in optimization and NE calculation, is not even convex. This result does not contradict Conjecture 1 as the dotted curve is the result of an inaccurate mathematical approximation that leads optimizations and NE calculations astray—for the latter case, selfish users acting distributedly can easily see it. The dotted curve is intentionally included in the figures with the purpose to show that inaccurate assumptions might not only lead to wrong decisions, but also hide important properties.

As a final remark, Fig. 4 shows that differences in terms of the maximum value for the PoA are not necessarily small and (i) when delays are uniform (and possibly tend to zero) the PoA tends to be smaller, whereas (ii) the shape of PoA curves does not match, with peaks occurring at different loads in the different cases. The latter tells that a correct incorporation of fixed latency is paramount to design a system meant for distributed optimization, as it would be otherwise impossible to predict the load at which the PoA will be maximum—hence more critical—and its associated value.

6   Conclusions

The analysis and optimization of computing task allocation in the edge-cloud continuum requires attention to system delays, whose effect is traditionally neglected. We have shown that incorporating such delays in the analysis of the latency experienced by end-users complicates the optimization and the NE identification with respect to the simpler cases commonly studied. Yet, the optimal allocation and the NE can be characterized analytically, and computed with the algorithms derived in this paper. Our analysis is general, and only requires the sojourn time of a task to be an increasing and convex function of the server load, which is a common property of non-fully-deterministic systems.

Our findings were validated through a real deployment spanning over a multiparty laboratory across different countries. We showed that (i) optimal configurations and selfish NEs can exhibit a strong dependency on relative differences between servers in terms of capacity and distance from the user, and (ii) distributed and selfish optimizations incur limited costs, unless the system is driven into deep saturation and the service time variance becomes unrealistically high.

The next steps of our work will consider multiserver systems, and queues with limited buffer, where losses can occur.

Acknowledgements. This work has been supported by the Project AEON-CPS (TSI-063000-2021-38), funded by the Ministry of Economic Affairs and Digital Transformation and the European Union NextGeneration-EU in the framework of the Spanish Recovery, Transformation and Resilience Plan, and the Italian National Recovery and Resilience Plan (NRRP), partnership on "Telecommunications of the Future" (PE0000001 - program "RESTART").

References

1. Bell, C.E., Stidham, S.: Individual versus social optimization in the allocation of customers to alternative servers. Manag. Sci. **29**, 831–839 (1983)
2. Braess, D.: Uber ein Paradoxon aus der Verkehrsplanung. Unternehmensforschung **12** (1969)
3. Cheng, Z., Gao, Z., Liwang, M., Huang, L., Du, X., Guizani, M.: Intelligent task offloading and energy allocation in the UAV-aided mobile edge-cloud continuum. IEEE Netw. **35**(5), 42–49 (2021)
4. Ding, Y., Li, K., Liu, C., Li, K.: A potential game theoretic approach to computation offloading strategy optimization in end-edge-cloud computing. IEEE Trans. Parallel Distrib. Syst. **33**(6), 1503–1519 (2021)
5. ETSI: TS 128 531 - V18.2.0 - 5G; Management and orchestration; Provisioning (Release 16). Technical specification, ETSI (2023)
6. Feldmann, R., Gairing, M., Lücking, T., Monien, B., Rode, M.: Selfish routing in non-cooperative networks: a survey. In: Rovan, B., Vojtáš, P. (eds.) MFCS 2003. LNCS, vol. 2747, pp. 21–45. Springer, Heidelberg (2003). https://doi.org/10.1007/978-3-540-45138-9_2
7. Garnaev, A., Trappe, W., Petropulu, A.: Equilibrium strategies for an OFDM network that might be under a jamming attack. In: Proceedings of IEEE CISS (2017)
8. Golang developers: Golang. https://go.dev
9. Hamilton, R., Iyengar, J., Swett, I., Wilk, A.: QUIC: A UDP-based secure and reliable transport for HTTP/2, tools.ietf.org/html/draft-tsvwg-quic-protocol-02
10. Haviv, M., Roughgarden, T.: The price of anarchy in an exponential multi-server. Op. Res. Lett. **35**(4), 421–426 (2007)
11. Koutsoupias, E., Papadimitriou, C.: Worst-case equilibria. Comput. Sci. Rev. **3**(2), 65–69 (2009)
12. Luo, Q., Hu, S., Li, C., Li, G., Shi, W.: Resource scheduling in edge computing: a survey. IEEE Commun. Surv. Tuts. **23**(4), 2131–2165 (2021)
13. Mancuso, V., Badia, L., Castagno, P., Sereno, M., Ajmone Marsan, M.: Efficiency of distributed selection of edge or cloud servers under latency constraints. In: Proceedings of IEEE MedComNet, pp. 158–166 (2023)
14. Mancuso, V., Castagno, P., Badia, L., Sereno, M., Ajmone Marsan, M.: Optimal Allocation of Tasks and Price of Anarchy of Distributed Optimization in Networked Computing Facilities. Technical report, arXiv:2404.05543 (2024). https://arxiv.org/abs/2404.05543
15. Milojicic, D.: The edge-to-cloud continuum. Computer **53**(11), 16–25 (2020)

16. Orda, A., Rom, R., Shimkin, N.: Competitive routing in multiuser communication networks. IEEE/ACM Trans. Netw. **1**(5), 510–521 (1993)
17. Pigou, A.C.: The Economics of Welfare. Macmillan, London (1920)
18. Rosen, J.B.: Existence and uniqueness of equilibrium points for concave n-person games. Econometrica J. Econometric Soc. 520–534 (1965)
19. Roughgarden, T.: Routing and the Price of Anarchy. MIT Press, Cambridge (2005)
20. Stidham, S.: Optimal Design of Queueing Systems. Chapman & Hall/CRC (2009)
21. Stidham, S.: The price of anarchy for a network of queues in heavy traffic. In: Pulat, P.S., Sarin, S.C., Uzsoy, R. (eds.) Essays in Production, Project Planning and Scheduling. ISORMS, vol. 200, pp. 91–121. Springer, Boston, MA (2014). https://doi.org/10.1007/978-1-4614-9056-2_5
22. Thai, M.T., Lin, Y.D., Lai, Y.C., Chien, H.T.: Workload and capacity optimization for cloud-edge computing systems with vertical and horizontal offloading. IEEE Trans. Netw. Service Manag. **17**(1), 227–238 (2020)
23. Wang, C.X., You, X., et al.: On the road to 6G: visions, requirements, key technologies and testbeds. IEEE Commun. Surv. Tuts. **25**(2), 905–974 (2023)
24. Wang, S., Zhao, Y., Xu, J., Yuan, J., Hsu, C.H.: Edge server placement in mobile edge computing. J. Parall. Distrib. Comput. **127**, 160–168 (2019)
25. Wang, T., Bauer, K., Forero, C., Goldberg, I.: Congestion-aware path selection for tor. In: Keromytis, A.D. (ed.) FC 2012. LNCS, vol. 7397, pp. 98–113. Springer, Heidelberg (2012). https://doi.org/10.1007/978-3-642-32946-3_9
26. Wu, T., Starobinski, D.: A comparative analysis of server selection in content replication networks. IEEE/ACM Trans. Netw. **16**(6), 1461–1474 (2008)
27. Yu, W., Cioffi, J.M.: Constant-power waterfilling: performance bound and low-complexity implementation. IEEE Trans. Commun. **54**(1), 23–28 (2006)
28. Zhang, J., Letaief, K.B.: Mobile edge intelligence and computing for the Internet of vehicles. Proc. IEEE **108**(2), 246–261 (2019)

Revenue Management for Parallel Services with Fully Observable Queues

Caitlin Vanden Bussche$^{(\boxtimes)}$![ORCID], Arnaud Devos ![ORCID], Sabine Wittevrongel ![ORCID], and Dieter Fiems ![ORCID]

Department of Telecommunications and Information Processing, Ghent University, Ghent, Belgium
Caitlin.VandenBussche@UGent.be

Abstract. This paper explores revenue optimisation in a queueing system with two observable parallel queues, Poisson arrivals and rational customers. Customers pay a predetermined fee which depends on the queue they wish to join. This fee is a parameter that the provider can set to her liking. While a higher fee leads to more profit per customer, it also incentivises customers to opt for the less costly queue. We consider both a scenario in which customers always select a queue and one in which customers are allowed to balk. In the former case, a power series approximation method is accelerated by Wynn's epsilon method. In the latter case, the queueing system constitutes a finite quasi-birth-death-type Markov chain and we rely on matrix-analytic techniques.

Keywords: Revenue Optimisation · Queueing Game · Join the Shortest Queue

1 Introduction

Revenue management involves optimising a service provider's revenue by adjusting pricing based on customer behaviour [1]. Our strategy involves offering varying service levels in parallel at different price points, empowering customers to select the option that best aligns with their preferences. By introducing differentiated fees, customers are presented with a choice between quicker, albeit pricier, service or more economical, albeit slower, service. This paper explores the optimisation of the provider's revenue by strategically selecting the fees for accessing one of two parallel queues.

Offering priority service access to premium customers is an obvious way to offer differentiated services and therefore well investigated. Gurvich et al. [2] compare a priority scheme used by a revenue-maximising firm and one chosen by a social planner. Relative to the social planner, a revenue maximising firm may have too few (ultra-luxury strategy) or too many (mass-luxury strategy) high-priority customers. In [3], Chamberlain and Starobinski consider a two-class unobservable priority queue where customers are charged a fee when joining the premium class. They compare the maximum revenue under non-preemptive and

© The Author(s), under exclusive license to Springer Nature Switzerland AG 2025
A. Devos et al. (Eds.): ASMTA 2024, LNCS 14826, pp. 51–66, 2025.
https://doi.org/10.1007/978-3-031-70753-7_4

preemptive-resume policies and show the preemptive-resume policy is always preferred by the service provider. Cao et al. [4] consider revenue maximisation for the non-observable $M/G/1$ non-preemptive priority queue, where customers are heterogeneous in their evaluation of the delay. This heterogeneity is introduced by assuming that each customer randomly draws its delay cost rate from a prescribed distribution. Service differentiation can also be obtained by non-strict priority disciplines. Revenue management for discriminatory processor sharing queues (DPS) is investigated in [5]. It is shown that an implementation of DPS is preferred over a strict priority discipline if the customers are allowed to balk. If this is not the case, the optimal DPS discipline is a strict priority discipline, even if the customers are heterogeneous in their evaluation of the waiting time. In all these priority models, customers choose strategically, in the sense that they choose the option which yields the largest expected utility. Hence, all these revenue management models combine a revenue optimisation problem with a queueing game.

In this paper, we focus on service differentiation by offering services in parallel. While it is well known that merging services is often beneficial (referred to in telecommunication networks as multiplexer gain), it is more than often practically inconvenient. For example, if the premium service offers an overall more premium experience, staff may require more training, the service may be offered at a more prime location, etc. The offer of parallel services means that customers now choose between different queues. Hence, our model can be seen as a generalisation of the Join-the-Shortest-Queue (JSQ) model which is widely studied. The JSQ model was first introduced and studied by Haight [6]. The author studies both the case where queueing customers are not permitted to switch to the other queue and the case where they are. In the latter case, the formulation becomes more simplified and explicit expressions can be obtained. Kingman [7] researched a similar JSQ system where customers are not permitted to change queues, thereby making the simplifying assumption that the queues are symmetrical. This assumption enables him to use generating functions to study the behaviour of the stationary solution. He proves that a state of statistical equilibrium is reached, as long as the arrival load does not exceed the server capacity. Later, Cohen wrote a short overview of studies on the symmetrical two-server queueing model, see [8] and presented the analytic solution for the asymmetrical two-server queueing model with negative exponentially distributed service times [9].

The present modelling assumptions generalise JSQ in that customers do not select a queue based on the length of the queues. Instead, they account for the total cost, which is the sum of a fee and the waiting time. The fee is a parameter that the provider can set to optimise her revenue. We solve this generalised JSQ model by a power series expansion approach, see e.g. [10] for an application of this approach for the join-the-shortest multi-server queueing model and [11,12] for more recent queueing applications. In general, the region of convergence of the series expansion is limited. We therefore rely on convergence acceleration techniques. Such techniques not only accelerate the convergence in the region of convergence, but also extend the region of convergence [13, 14]. We here particularly apply Wynn's ϵ-algorithm to improve convergence. Wynn's ϵ-algorithm [15] is an implementation of the Shank's transformation, the best all-purpose acceleration method according to Graves-Morris and Jenkins [16]. Under nonstringent assumptions on the sequences, the transformation converges faster to the limit of the original series in the region of convergence, and converges to the analytic continuation of diverging sequences [17].

The remainder of this paper is organised as follows. In the next section, our specific modelling assumptions and the queueing analysis are presented. We also illustrate the accuracy of our approach by a numerical example. The problem of revenue management for the queueing model at hand is then discussed in Sect. 3. Without the possibility of balking, we find that the revenue management problem is trivial if the arrival intensity exceeds one of the servers' capacity. We therefore also investigate a model where customers are allowed to balk in Sect. 4. We specifically discuss how balking affects revenue management. Finally, conclusions are drawn in Sect. 5.

2   Queueing Model and Analysis

We analyse a continuous-time Markovian queueing model, with two parallel queues with dedicated servers. The customer service times in the first and second queue constitute sequences of independent random variables, exponentially distributed with rates μ_1 and μ_2 for the first and second queue, respectively. If $\mu_1 = \mu_2$ we call the model symmetric, if $\mu_1 \neq \mu_2$ we have an asymmetric system. Customers arrive at the system according to a Poisson process with an arrival rate λ and choose between the queues after observing the state of the system.

The choice of queue depends on the expected value V_i that the service in queue i brings, the entry fee F_i for entering the ith queue and the expected waiting time W_i in the ith queue ($i \in \{1, 2\}$). In view of the assumptions, with m customers in the first and n in the second queue, the waiting times in the queues equal

$$W_1(m, n) = \frac{m}{\mu_1}, \quad W_2(m, n) = \frac{n}{\mu_2}.$$

Accounting for the service value, and the cost in terms of entry fees and waiting times, let U_i denote the utility of entering queue i,

$$U_i(m, n) = V_i - F_i - W_i(m, n).$$

Customers now select the queue that brings the highest utility. In case both queues offer the same utility, queue 2 is preferred over queue 1: queue 1 is selected if $U_1(m, n) > U_2(m, n)$ and queue 2 is selected if this is not the case. Note that in the expression above, the waiting times are expressed in time units. To simplify the notation, we did not add conversion factors to the different terms, meaning that V_i and F_i are also expressed in terms of time units.

Let $Q_i(t)$ denote the number of customers in the ith queue at time t, $i \in \{1, 2\}$. With the assumptions on the arrival and service processes above, as well as with the assumptions on how customers select one queue over the other, it is now easily verified that the process $\{(Q_1(t), Q_2(t)), t > 0\}$ constitutes a Markov process. This Markov process admits a limiting distribution, provided that the following stability condition holds,

$$\lambda < \mu_1 + \mu_2 .$$

Let $\{\pi(m, n); n, m \in \mathbb{N}\}$ denote the limiting distribution,

$$\pi(m, n) = \lim_{t \to \infty} \mathsf{P}[Q_1(t) = m, Q_2(t) = n] .$$

We now have the following balance equations for our model,

$$\pi(m, n) \left(\mu_1 \mathbb{1}_{\{m>0\}} + \mu_2 \mathbb{1}_{\{n>0\}} + \lambda \right) = \pi(m+1, n)\mu_1 + \pi(m, n+1)\mu_2$$
$$+ \pi(m-1, n)\lambda \mathbb{1}_{\{U_1(m-1,n)>U_2(m-1,n)\}}$$
$$+ \pi(m, n-1)\lambda \mathbb{1}_{\{U_1(m,n-1)\leq U_2(m,n-1)\}} ,$$

for $m, n \in \mathbb{N}$. Here $\mathbb{1}_{\{\cdot\}}$ denotes the indicator function which equals 1 if its argument is true and 0 if it is not. Moreover, we set $\pi(m, n) = 0$ for $(m, n) \notin \mathbb{N}^2$.

In line with the literature on JSQ systems, the balance equations are not easily solved. We therefore rely on a power series approximation approach to solve the set of balance equations, in line with Blanc's approach to JSQ systems [10]. To this end, we introduce the following power series expansion for the stationary probability $\pi(m, n)$,

$$\pi(m, n) = \lambda^{m+n} \sum_{k=0}^{\infty} \lambda^k u_k(m, n) . \tag{1}$$

Note that the expression above implicitly assumes that the first $m + n$ terms in the series expansion are zero. This is indeed the case and follows from the so-called N-events rule: the Nth order term in the series expansion of a state is non-zero only if this state can be reached in at most N λ-transitions. In this case, one needs at least $n + m$ arrivals to reach state (m, n).

As indicated in [10], the advantage of the power series expansion approach over others is that the coefficients $u_k(m, n)$ can be obtained recursively. Once these coefficients are available, the stationary probabilities $\pi(m, n)$ are easily calculated for different values of λ. To obtain the coefficients $u_k(m, n)$ we first substitute the stationary probabilities by their power series expansions in the balance equations. If we then equate the coefficients of the terms in equal powers of λ on both sides of the balance equations, we find

$$u_k(m, n) = \Big(-u_{k-1}(m, n) + u_{k-1}(m+1, n)\mu_1 + u_{k-1}(m, n+1)\mu_2$$
$$+ u_k(m-1, n)\mathbb{1}_{\{U_1(m-1,n)>U_2(m-1,n)\}}$$
$$+ u_k(m, n-1)\mathbb{1}_{\{U_1(m,n-1)\leq U_2(m,n-1)\}} \Big) \times \Big(\mu_1 \mathbb{1}_{\{m>0\}} + \mu_2 \mathbb{1}_{\{n>0\}} \Big)^{-1} ,$$

for $k \in \mathbb{N}$ and $(m, n) \in \mathbb{N}^2 \setminus (0, 0)$, with the agreement that $u_k(m, n) = 0$ if the index k or any of the arguments m or n are negative.

Clearly, the expression above can be used to recursively calculate $u_k(m, n)$ for all $(m, n) \neq (0, 0)$ and all k. For $(m, n) = (0, 0)$ we find the following equations which follow in a similar way from the normalisation condition of the stationary distribution,

$$u_0(0, 0) = 1,$$

and

$$u_k(0, 0) = - \sum_{n_1=0}^{k} \sum_{n_2=0}^{k-n_1} u_{k-(n_1+n_2)}(n_1, n_2) \mathbb{1}_{\{n_1+n_2 \neq 0\}}.$$

Once these terms have been calculated, we can approximate the stationary probabilities by truncating the summation in equation (1). In turn, the approximate expressions of the stationary probabilities can be used to approximate various performance measures of interest. For revenue management, the fraction of customers p that opt for queue 1 is a key performance measure. By PASTA (Poisson arrivals see time averages), $\pi(m, n)$ is the probability that an arriving customer finds m customers in the first queue and n in the second upon arrival. Hence, we can express p as follows,

$$p = \sum_{m=0}^{\infty} \sum_{n=0}^{\infty} \pi(m, n) \mathbb{1}_{\{U_1(m,n) > U_2(m,n)\}}.$$

By plugging the series expansion (1) in the expression above, we find the following series expansion for p,

$$p = \sum_{\ell=0}^{\infty} a_\ell \lambda^\ell,$$

with

$$a_\ell = \sum_{m=0}^{\ell} \sum_{n=0}^{\ell-m} u_{\ell-m-n}(m, n) \mathbb{1}_{\{U_1(m,n) > U_2(m,n)\}}.$$

Similar expressions can also be obtained for the mean number of customers in queue 1 and queue 2,

$$\overline{Q}_1 = \sum_{\ell=0}^{\infty} b_{1,\ell} \lambda^\ell, \quad \overline{Q}_2 = \sum_{\ell=0}^{\infty} b_{2,\ell} \lambda^\ell,$$

with

$$b_{1,\ell} = \sum_{m=0}^{\ell} \sum_{n=0}^{\ell-m} u_{\ell-m-n}(m, n)m, \quad b_{2,\ell} = \sum_{m=0}^{\ell} \sum_{n=0}^{\ell-m} u_{\ell-m-n}(m, n)n.$$

The numerical complexity of calculating the kth term in the series expansion is $O(k^2)$. Hence, one can easily calculate many terms in the expansion.

Higher order expansions lead to increasingly accurate approximations of the performance of interest in the region of convergence of the series expansion. Unfortunately, the radius of convergence of the series expansion may be limited due to the presence of negative real singularities or complex singularities. We can however both improve convergence in the region of convergence and extend the region of convergence by means of Wynn's epsilon algorithm [15].

Let s_ℓ denote the coefficients in the series expansion of a performance measure of interest. We have $s_\ell = a_\ell$ for the fraction p and $s_\ell = b_{i,\ell}$ for the mean queue content $\overline{Q}_i$, $i \in \{1,2\}$. Wynn's epsilon algorithm consists of the following recursive scheme,

$$\epsilon_{\kappa+1}(S_m) = \epsilon_{\kappa-1}(S_{m+1}) + [\epsilon_\kappa(S_{m+1}) - \epsilon_\kappa(S_m)]^{-1} ,$$

for $m = 0, 1, \ldots$ and $\kappa = 0, 1, \ldots$, where we choose as initial values

$$\epsilon_{-1}(S_m) = 0, \quad \epsilon_0(S_m) = S_m ,$$

with $S_m = \sum_{\ell=0}^{m} s_\ell \lambda^\ell$ the mth partial sum of the power series. The even sequences $\{\epsilon_{2\kappa}(S_m), \ \kappa = 0, 1, \ldots\}$ are sequences which typically converge faster to the desired result than the partial sums $\{S_m, \ m = 0, 1, \ldots\}$. The odd sequences $\{\epsilon_{2\kappa+1}(S_m), \ \kappa = 0, 1, \ldots\}$ are just intermediate steps. Practically, if we start with $2M + 1$ terms S_m ($m = 0, 1, \ldots, 2M$), we can calculate $\epsilon_{2M}(S_0)$.

Wynn's epsilon algorithm is fundamentally an extrapolation method where our sequence is transformed into a new sequence, which under some conditions converges faster to the same limit. The approximation is related to the well known Padé approximants, see [17] for a more detailed explanation.

To illustrate the accuracy of our approach, we compare the (approximate) analytical results with results obtained by stochastic simulation. Figure 1 depicts the fraction of customers p that opt for the first queue as a function of the arrival rate λ. The curves represent analytical results, the markers represent simulation results. Both queues offer the same customer experience ($V_1 = V_2$) such that the difference in fees $\Delta F = F_2 - F_1$ and the waiting times determine the preference of one queue over the other. Figure 1a depicts p for a symmetric system ($\mu_1 = \mu_2$), while Fig. 1b depicts a system where the larger fee applies to a faster server ($\mu_1 = 1, \mu_2 = 2$). In both figures, different values of ΔF are assumed as indicated. For $\lambda = 0$ all customers opt for the cheaper queue. In the absence of convergence, the choice is entirely determined by the fee. When λ increases, more customers are prepared to pay the extra fee. When the difference in the fees increases, more customers opt for the first queue, which is not unexpected. Finally, for λ approaching $\mu_1 + \mu_2$, p converges to $\mu_1/(\mu_1 + \mu_2)$. For very high load, one cannot deviate considerably from this traffic division, to ensure that both queues are stable.

3 Revenue Management

So far, we have focused on the strategies followed by the different customers. We now focus on the service provider's strategy. The provider optimises revenue by

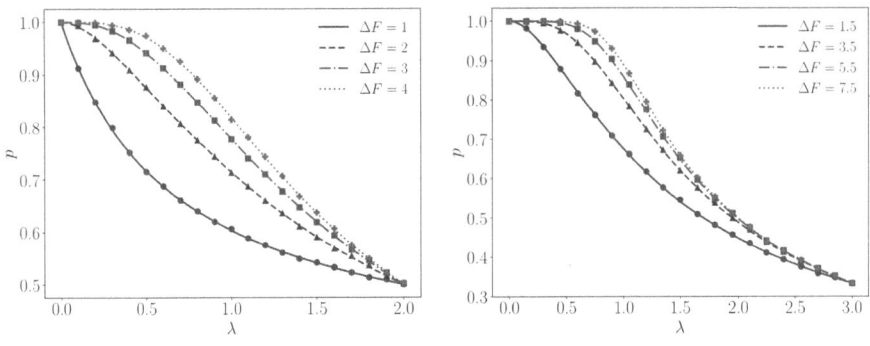

(a) Symmetrical case: $\mu_1 = \mu_2 = 1$ (b) Asymmetrical case: $\mu_1 = 1$ and $\mu_2 = 2$

Fig. 1. Fraction p of customers that opt for queue 1 vs. the arrival rate λ for $V_1 = V_2$ and for different ΔF as indicated. The curves are obtained by the series expansion approach, the markers by stochastic simulation.

fine-tuning the entry fees for the different queues. As all customers choose either one or the other, the obvious optimum follows from selecting very high entry fees for both queues. This is not very realistic though, as by increasing the fees it becomes more likely that customers do not join the queues, an effect which is not captured by the present modelling assumptions and which will be explored in the next section.

Nevertheless, also in the present setting, revenue management can be studied by focusing on the difference between the fees and on the corresponding differences in revenue. To this end, let C_i denote the cost for serving a customer in the ith queue. As before, let $\Delta F = F_2 - F_1$ denote the difference between entry fees, and let $\Delta C = C_2 - C_1$ denote the difference in cost per customer. The revenue per time unit then equals,

$$R = R_1 + R_2 = \lambda p(F_1 - C_1) + \lambda(1 - p)(F_2 - C_2) = \lambda p(\Delta C - \Delta F) + \lambda(F_2 - C_2).$$

Here p is the proportion of customers that opt for the first queue, which was calculated in the preceding section and which depends on ΔF (but not on ΔC).

We now study the evolution of R as a function of ΔF. To do so, it is instructive to return to the balance equations. Let $\Delta V = V_2 - V_1$ denote the difference in expected service value and let $\mathcal{D}$ denote the set

$$\mathcal{D} = \left\{ \Delta V + \frac{m}{\mu_1} - \frac{n}{\mu_2}; m, n \in \mathbb{N} \right\}. \tag{2}$$

This is the set of points at which the balance equations change while increasing or decreasing ΔF. In a neighbourhood of any intermediate ΔF value, the balance equations remain constant as the indicator functions in the balance equations evaluate to the same value. There is such a change for state (m, n) when the following equality holds,

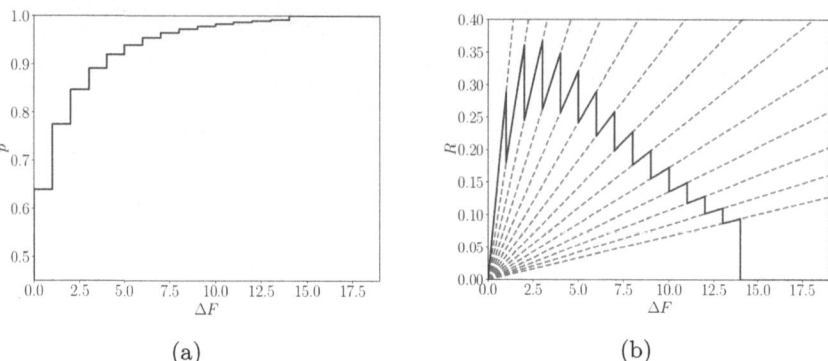

Fig. 2. Fraction p and revenue R vs. the difference ΔF, for $\lambda = 0.8$ and $\mu_1 = \mu_2 = 1$.

$$U_1(m,n) = U_2(m,n) \iff V_1 - F_1 - \frac{m}{\mu_1} = V_2 - F_2 - \frac{n}{\mu_2}.$$

Equivalently, we have a change for

$$\Delta F = \Delta V + \frac{m}{\mu_1} - \frac{n}{\mu_2}.$$

As the balance equations do not change, neither does the fraction of customers p. Hence, p is a piecewise constant function of ΔF. Moreover, p is an increasing function of ΔF, as more customers opt for the first queue if its entry fee decreases. The revenue R is a piecewise linear function, with discontinuities for $\Delta F \in \mathcal{D}$. The evolution between two points of $\mathcal{D}$ is an affine function with slope λp, intersecting the real axis for $\Delta F = \Delta C$. Finally note that if we have equal service times ($\mu_1 = \mu_2 = \mu$), the set $\mathcal{D}$ simplifies to

$$\mathcal{D} = \left\{ \Delta V + \frac{m}{\mu}; m \in \mathbb{Z} \right\}.$$

We now study revenue management by some numerical examples. In Fig. 2 we consider a symmetrical model with equal customer values, $V_1 = V_2$, arrival rate $\lambda = 0.8$ and service rates $\mu_1 = \mu_2 = 1$. With these assumptions the set of discontinuities $\mathcal{D}$ coincides with the natural numbers. Moreover, each queue can accommodate all customers ($\lambda < \mu_1 = \mu_2$). Figure 2a depicts the fraction of customers that opt for the first queue, p, as a function of the difference in fees, ΔF. As expected, we observe that an increase of ΔF leads to more customers opting for the first queue, and p increases till all customers opt for this queue ($p = 1$). Figure 2b depicts the total revenue R of the provider as a function of the difference in fees ΔF, assuming $F_1 = 0$. Since the revenue R is a function of p, there is also a jump discontinuity at each point of $\mathcal{D}$. Between these jumps, the revenue increases according to a linear function through the origin $(0, 0)$. We added the grey dashed lines to illustrate this.

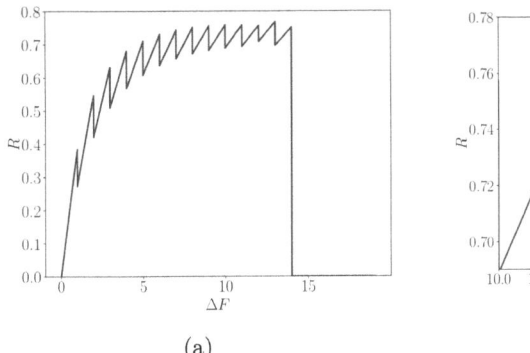
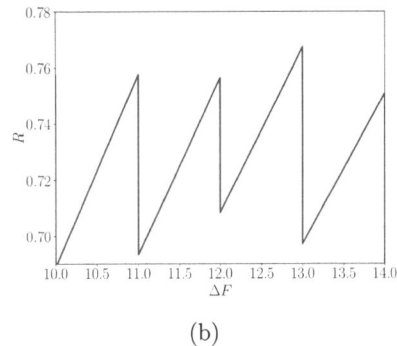

(a)   (b)

Fig. 3. Revenue R vs. the difference ΔF, for $\lambda \approx 0.978204$.

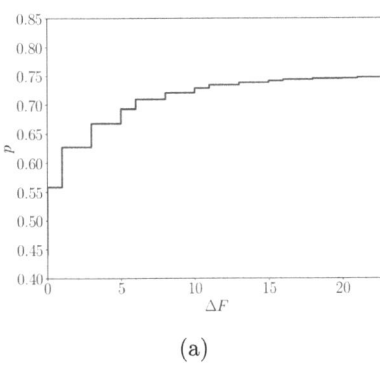
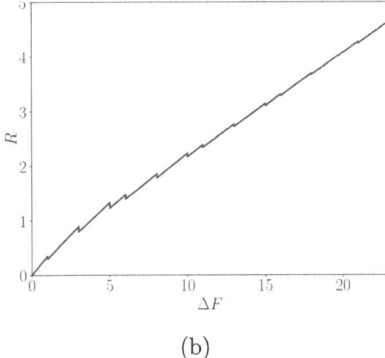

(a)   (b)

Fig. 4. Fraction p and revenue R vs. the difference ΔF for $\lambda = 0.8$ and $\mu_1 = \mu_2 = 0.6$.

Clearly, the revenue is not a convex function of ΔF. The optimisation problem however simplifies as the optimal revenue is obtained at a discontinuity point. As a result, it suffices to consider only these points. The evolution of the revenue in Fig. 2b suggests that if we only consider the discontinuity points, we have a property somewhat similar to convexity. The revenue at the jump points initially increases and then decreases to zero. This trend appears consistent across many symmetrical models where $\lambda < \mu_1 = \mu_2$, suggesting it might be a general pattern. However, we can construct a counterexample: the revenue for $\lambda \approx 0.978204$, $\mu_1 = \mu_2$ and $V_1 = V_2$ is depicted in Fig. 3. Again focusing on the discontinuities, there is an increase in revenue after an initial decrease. As the revenue is limited, Fig. 3b zooms in on these discontinuity points.

In Fig. 4, we consider a model similar to the previous one, except that now $\lambda > \mu_1 = \mu_2$. In this model, directing all customers to a single queue would render it unstable as its system content would go to infinity. Figure 4a depicts the fraction p as a function of the difference in fees, ΔF. If ΔF increases, more customers opt for the first queue, causing p to increase in the limit to $\frac{\mu_1}{\lambda} < 1$,

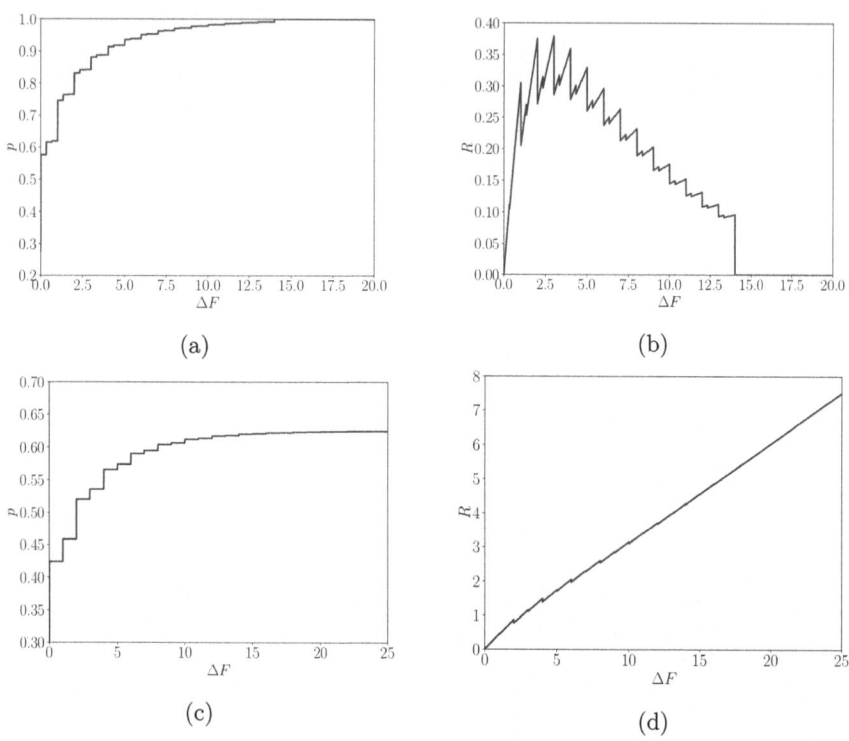

Fig. 5. Fraction p and revenue R vs. the difference ΔF for $\lambda = 0.8$. For (a) and (b), we have $\mu_1 = 1$ and $\mu_2 = 1.5$. For (c) and (d), we have $\mu_1 = 0.5$ and $\mu_2 = 1$.

the point where the queue load reaches 1. As a result, there always remains a fraction $1 - p$ of the customers who choose the second, pricier queue. Revenue is therefore optimal for $\Delta F \to \infty$ as depicted in Fig. 4b.

We now turn to some examples where we make the second queue more appealing to customers by offering faster service. Figure 5 depicts the fraction p and revenue R for two asymmetrical models with equal customer experience, $V_1 = V_2$ and arrival rate $\lambda = 0.8$. In Figs. 5a and 5b the service rates are $\mu_1 = 1$ and $\mu_2 = 1.5$. In this example, we have that $\lambda < \min(\mu_1, \mu_2)$, so each queue can accommodate all customers. The behaviour of the system is similar to the symmetric case, but there are more jump discontinuities which no longer coincide with the natural numbers, see equation (2). In Figs. 5c and 5d the service rates are $\mu_1 = 0.5$ and $\mu_2 = 1$. The difference with the previous example is that $\lambda > \mu_1$, so not all customers can be served by the first queue, as this would make the system unstable. As depicted in Fig. 5c, the fraction p now converges in the limit to $\frac{\mu_1}{\lambda} = 0.625$. This is the point where the load of the first queue reaches 1, and the first queue becomes unstable. Once again this leads to a revenue that continues to rise with increasing ΔF, as illustrated in Fig. 5d.

4   Revenue Management with Balking

Recall that $U_i(m,n)$ denotes the utility a customer receives when entering queue i when there are m customers in queue 1 and n customers in queue 2 upon arrival. In contrast to the preceding sections, we now assume that a customer balks whenever her utility is negative. The balance equations then read

$$\pi(m,n)\left(\mu_1\mathbb{1}_{\{m>0\}} + \mu_2\mathbb{1}_{\{n>0\}} + \lambda\mathbb{1}_{\{\max(U_1(m,n),U_2(m,n))\geq 0\}}\right) =$$
$$\pi(m+1,n)\mu_1 + \pi(m,n+1)\mu_2$$
$$+ \pi(m-1,n)\lambda\mathbb{1}_{\{U_1(m-1,n)>U_2(m-1,n),U_1(m-1,n)\geq 0\}}$$
$$+ \pi(m,n-1)\lambda\mathbb{1}_{\{U_2(m,n-1)\geq U_1(m,n-1),U_2(m,n-1)\geq 0\}} \cdot$$

In contrast to the scenario without balking, the state space of the Markov chain is finite. Indeed, there are no customer arrivals in the first and second queue for $m \geq M$ and $n \geq N$ respectively, with

$$M = \lfloor \mu_1(V_1 - F_1)\rfloor + 1, \quad N = \lfloor \mu_2(V_2 - F_2)\rfloor + 1.$$

As both queues are finite, and the queue content changes by at most one with arrivals and departures, the corresponding Markov chain admits a level-dependent quasi-birth-death (QBD) structure. Indeed, if we let the first queue size correspond to the level and the second queue size to the phase of the QBD, we easily see that there are only transitions to adjacent levels. Moreover, phase transitions do not alter the level, and level transitions do not alter the phase.

Let $\boldsymbol{\pi}(m)$ denote the row vector with entries $\pi(m,n)$, $n = 0,\ldots,N$. The balance equations can then be rewritten in matrix notation as follows,

$$\boldsymbol{\pi}(0)L(0) + \mu_1\boldsymbol{\pi}(1) = 0$$
$$\boldsymbol{\pi}(m-1)F(m-1) + \boldsymbol{\pi}(m)L(m) + \mu_1\boldsymbol{\pi}(m+1) = 0$$
$$\boldsymbol{\pi}(M-1)F(m-1) + \boldsymbol{\pi}(M)L(M) = 0.$$

Here the local (phase) transition matrix L and the forward (level) transition matrix F are defined by

$$L(m) = \mu_2\mathbf{e}_0'\mathbf{e}_0 - \left(\mu_1\mathbb{1}_{\{m>0\}} + \mu_2\right)I - \lambda\sum_{n=0}^{N}\mathbf{e}_n'\mathbf{e}_n\mathbb{1}_{\{\max(U_1(m,n),U_2(m,n))\geq 0\}}$$
$$+ \mu_2\sum_{n=0}^{N}\mathbf{e}_{n+1}'\mathbf{e}_n + \lambda\sum_{n=0}^{N}\mathbf{e}_{n-1}'\mathbf{e}_n\mathbb{1}_{\{U_2(m,n-1)\geq U_1(m,n-1),U_2(m,n-1)\geq 0\}},$$

and

$$F(m) = \sum_{n=0}^{N}\mathbf{e}_n'\mathbf{e}_n\lambda\mathbb{1}_{\{U_1(m,n)>U_2(m,n),U_1(m,n)\geq 0\}},$$

where I is the identity matrix and $\mathbf{e}_n$ is a row vector of zeroes, with the nth entry equal to 1 if $0 \leq n \leq N$ and $\mathbf{x}'$ denotes the transpose of $\mathbf{x}$. Having defined the block structure of the QBD, one easily finds the stationary solution of the QBD, e.g. by linear level reduction.

The revenue per time unit equals

$$R = \lambda p_1 (F_1 - C_1) + \lambda p_2 (F_2 - C_2),$$

with p_i the fraction of customers that opt for the ith queue,

$$p_1 = \sum_{m=0}^{M} \sum_{n=0}^{N} \pi(m,n) \mathbb{1}_{\{U_1(m,n) > U_2(m,n), U_1(m,n) \geq 0\}},$$

$$p_2 = \sum_{m=0}^{M} \sum_{n=0}^{N} \pi(m,n) \mathbb{1}_{\{U_2(m,n) \geq U_1(m,n), U_2(m,n) \geq 0\}}.$$

The fraction of customers that choose to balk is then $1 - p_1 - p_2$.

We now study revenue management by some numerical examples. Figure 6 depicts the customer choice as a function of the arrival rate λ. Figures 6a and 6b depict p_1, the fraction of customers that choose queue 1. Figures 6c and 6d depict p_2, the fraction of customers that opt for queue 2, while Figs. 6e and 6f depict $1 - p_1 - p_2$, the fraction of customers that balk.

The left panes in Fig. 6 consider a symmetrical model with $\mu_1 = \mu_2 = 1$. Both queues offer the same customer values, $V_1 = V_2 = 50$, so the difference in fees ΔF and the waiting times determine the customer behaviour. Different values of ΔF are assumed as indicated. We see similar behaviour as in the model without balking. For $\lambda = 0$, all customers opt for the cheaper queue. When λ increases, more customers are prepared to pay the extra fee, due to increasing waiting times in the cheaper queue. When the fee differs more, for larger ΔF, more customers opt for the first queue. If $\lambda < \mu_1 + \mu_2 = 2$, both queues can accommodate all customers and nobody balks as the customers' service value is high enough ($V_1 = V_2 = 50$). In the case where $\lambda > \mu_1 + \mu_2 = 2$, the queues cannot accommodate all customers. As a result a fraction of the customers balks because otherwise their waiting times and fees exceed their service value. From that point on, the fractions are the same for the different values of ΔF considered.

On the right side in Fig. 6 we consider an asymmetrical model (with $\mu_1 = 0.5$ and $\mu_2 = 1$), and focus on different customer values $V_1 = V_2$ as indicated. The fees are constant, $\Delta F = 2$ and $F_1 = 2$. The expected service values and the waiting times determine the choice of the customers. Again for $\lambda = 0$, all customers choose the cheaper queue. If the customer service values, V_1 and V_2, are small, customers tend to balk more, even at low values of λ. If these values increase, customers only start balking when $\lambda > \mu_1 + \mu_2$.

A provider can only try to maximise her revenue by changing the fees. Figure 7 depicts two models with equal expected service values $V_1 = V_2 = 10$ and where the first queue is free of charge, i.e. $F_1 = 0$. In Figs. 7a and 7b we consider a symmetrical model with arrival rate $\lambda = 1.5$ and service rates $\mu_1 = \mu_2 = 1$.

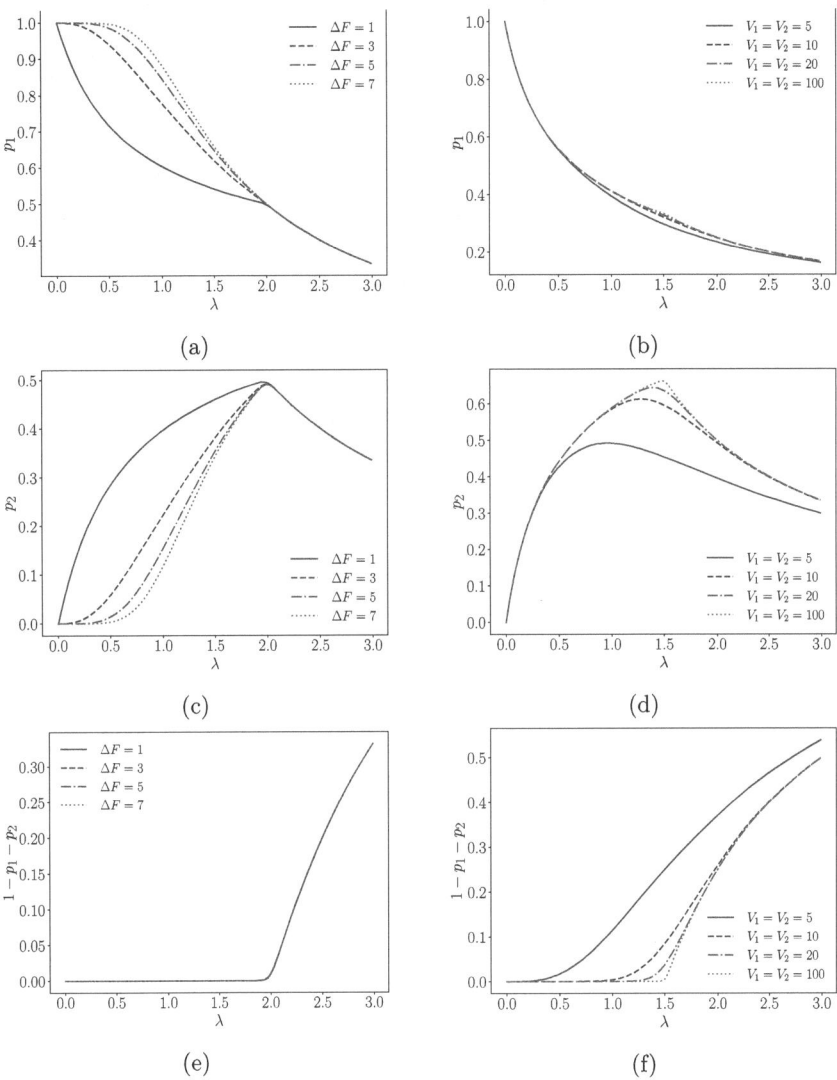

Fig. 6. Fractions p_1, p_2 and $1 - p_1 - p_2$ of customers vs. the arrival rate λ. The left panes show results for a symmetric model ($\mu_1 = \mu_2 = 1$, $V_1 = V_2 = 50$). The right panes assume an asymmetric model ($\mu_1 = 0.5$, $\mu_2 = 1$, $F_1 = 2$, $F_2 = 4$).

Figure 7a displays the choice customers make in function of the difference in fee ΔF (which is equal to F_2 in our case). Similarly as in the case without balking, we observe a jump discontinuity at the points in the set $\mathcal{D}$ which coincides with the natural numbers. Between these discontinuity points, the choice of the customers does not change. The fraction p_1 of customers that opt for the first queue will approach $\frac{\mu_1}{\lambda} \approx 0.66$ if ΔF increases. The fraction p_2 of customers that

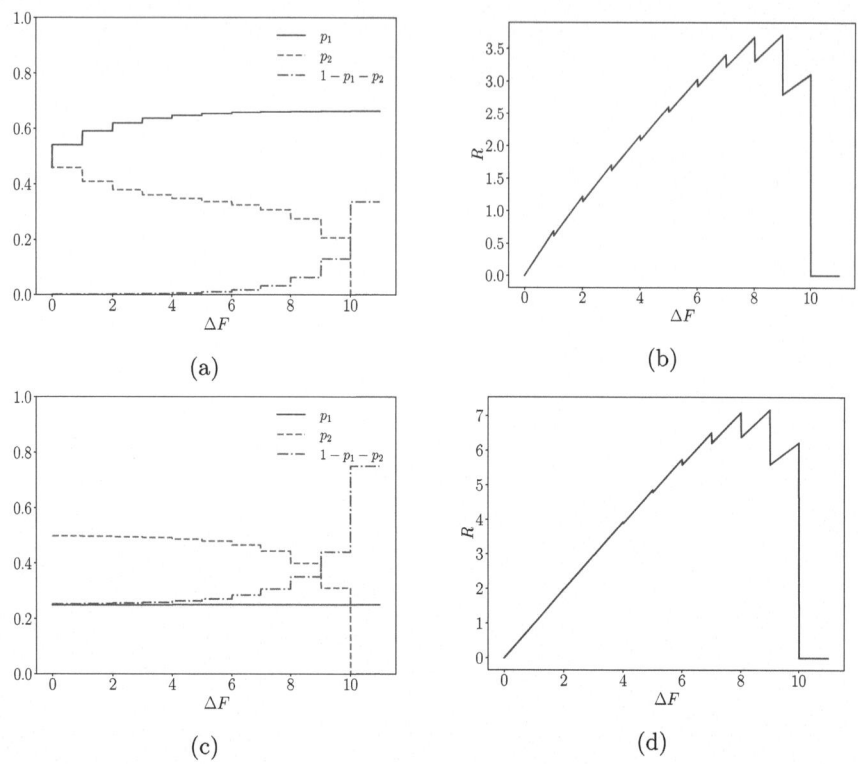

Fig. 7. Customer choice and revenue vs. the difference ΔF for $V_1 = V_2 = 10$ and $F_1 = 0$. For (a) and (b), we have $\lambda = 1.5$ and $\mu_1 = \mu_2 = 1$. For (c) and (d), we have $\lambda = 2$, $\mu_1 = 0.5$ and $\mu_2 = 1$.

opt for the second queue will tend to zero as ΔF increases. This is because the fraction of customers that balk will increase if ΔF increases, as more and more customers will find joining too expensive. Figure 7b displays the corresponding revenue of the provider. There are also discontinuities at the points of the set $\mathcal{D}$. The optimal revenue is reached at a point of this set $\mathcal{D}$ and can thus easily be found.

In Figs. 7c and 7d we consider an asymmetrical model with arrival rate $\lambda = 2$ and service rates $\mu_1 = 0.5$ and $\mu_2 = 1$. We see similar behaviour as in the previous figures. Figure 7c depicts the choice customers make in function of the difference in fees ΔF. Not all customers can be accommodated by our system, so all queues work at their limit, with a fraction $\frac{\mu_i}{\lambda}$ of the customers being served in queue i, if the second queue is also free, i.e. $\Delta F = F_2 = 0$. There is also a fraction of at least $\frac{\lambda - \mu_1 - \mu_2}{\lambda}$ customers that balk. If ΔF increases, this fraction that balks increases and the fraction p_2 decreases because customers will find the queue too expensive. We see that the first queue keeps working at its limit as it is not affected by the fee in the second queue. The maximum revenue, depicted in Fig. 7d, can again be easily found by limiting the search to points in $\mathcal{D}$.

5   Conclusions

In this paper, we investigated a queueing system with two observable parallel queues and dedicated servers, where customers decide which queue to join based on the entering fees, the waiting times and their expected service value. Analytical solutions are not feasible for this queueing model due to its complexity, so we used a power series expansion approach. To improve the convergence of the series expansion, Wynn's epsilon algorithm was applied and the accuracy was verified by simulation. If the fee of a certain queue increases, the fraction of customers that will opt for the cheaper queue will increase to that queue's maximal capability. We further showed that for finding the maximum revenue one only needs to look at a well specified set of discontinuity points. As our model does not allow for balking, the provider can set entry fees at will. As this is not very realistic, we also considered a model where customers can balk if the cost of joining either queue exceeds their expected service value and illustrated revenue management by a numerical example.

References

1. Ng, F., Rouse, P., Harrison, J.: Classifying revenue management: a taxonomy to assess business practice. Decis. Sci. **48**, 489–522 (2017)
2. Gurvich, I., Lariviere, M.A., Ozkan, C.: Coverage, coarseness, and classification: determinants of social efficiency in priority queues. Manage. Sci. **65**(3), 1061–1075 (2019)
3. Chamberlain, J., Starobinski, D.: Strategic revenue management of preemptive versus non-preemptive queues. Oper. Res. Lett. **49**(2), 184–187 (2021)
4. Cao, P., Wang, Y., Xie, J.: Priority service pricing with heterogeneous customers: impact of delay cost distribution. Prod. Oper. Manag. **28**(11), 2854–2876 (2019)
5. Fiems, D.: Strategic revenue management for discriminatory processor sharing queues. LNCS, vol. 14231, pp. 1–17 (2023). https://doi.org/10.1007/978-3-031-43185-2_1
6. Haight, F.A.: Two queues in parallel. Biometrika **45**(3/4), 401–410 (1958)
7. Kingman, J.F.C.: Two similar queues in parallel. Ann. Math. Stat. **32**(4), 1314–1323 (1961)
8. Cohen, J.W.: On the analysis of the symmetric shortest queue. In: Quantitative Methods in Parallel Systems, pp. 141–152. Springer (1995)
9. Cohen, J.W.: Analysis of the asymmetrical shortest two-server queueing model. J. Appl. Math. Stoch. Anal. **11**(2), 115–162 (1998)
10. Blanc, J.P.C.: A note on waiting times in systems with queues in parallel. J. Appl. Probab. **24**(2), 540–546 (1987)
11. De Cuypere, E., De Turck, K., Wittevrongel, S., Fiems, D.: Opinion propagation in bounded medium-sized populations. Perform. Eval. **99–100**, 1–15 (2016)
12. Fiems, D., Phung-Duc, T.: Light-traffic analysis of random access systems without collisions. Ann. Oper. Res. **277**, 311–327 (2019)

13. Weniger, E.J.: Nonlinear sequence transformations for the acceleration of conver-
    gence and the summation of divergent series. *Computer Physics Reports*, pages
    189–371, 1989
14. Brezinski, C., Redivo-Zaglia, M.: Extrapolation Methods: Theory and Practice.
    North-Holland (1991)
15. Wynn, P.: The epsilon algorithm and operational formulas of numerical analysis.
    Math. Comput. **15**(74), 151–158 (1961)
16. Graves-Morris, P.R., Jenkins, C.D.: Vector-valued rational interpolants iii. Constr.
    Approx. **2**, 263–289 (1986)
17. Brezinski, C., Redivo-Zaglia, M.: A survey of shanks' extrapolation methods and
    their applications. Comput. Math. Math. Phys. **61**(5), 699–718 (2021)

Deep Reinforcement Learning for Weakly Coupled MDP's with Continuous Actions

Francisco Robledo[1,2(✉)] [ID], Urtzi Ayesta[1,3,4] [ID], and Konstantin Avrachenkov[5] [ID]

[1] UPV/EHU, Univ. of the Basque Country, 20018 Donostia, Spain
frrobledo96@gmail.com
[2] UPPA, Université de Pau et des Pays de l'Adour, 64000 Pau, France
[3] IRIT, Université de Toulouse, CNRS, Toulouse INP, UT3, Toulouse, France
urtzi.ayesta@irit.fr
[4] IKERBASQUE - Basque Foundation for Science, 48011 Bilbao, Spain
[5] INRIA, Sophia, Antipolis, France
k.avrachenkov@inria.fr

Abstract. This paper introduces the Lagrange Policy for Continuous Actions (LPCA), a reinforcement learning algorithm specifically designed for weakly coupled MDP problems with continuous action spaces. LPCA addresses the challenge of resource constraints dependent on continuous actions by introducing a Lagrange relaxation of the weakly coupled MDP problem within a neural network framework for Q-value computation. This approach effectively decouples the MDP, enabling efficient policy learning in resource-constrained environments. We present two variations of LPCA: LPCA-DE, which utilizes differential evolution for global optimization, and LPCA-Greedy, a method that incrementally and greadily selects actions based on Q-value gradients. Comparative analysis against other state-of-the-art techniques across various settings highlight LPCA's robustness and efficiency in managing resource allocation while maximizing rewards.

Keywords: Reinforcement Learning · Weakly Coupled MDP · Continuous Actions · Lagrange Policy · Neural Networks · Differential Evolution · Resource Allocation · Policy Optimization

1 Introduction

The exploration of optimal decision-making under uncertainty is a fundamental problem [17], with significant implications in diverse fields such as telecommunications, finance, robotics, and healthcare. At the heart of this exploration lies the restless multi-armed bandit (RMAB) problem, an extension of the classical multi-armed bandit framework [6] to scenarios where arms evolve independently of the player's actions. Introduced by [21], the RMAB problem highlights the challenge of allocating limited resources among competing projects or processes in a state of continuous change. Recently, many studies have focused on neural network approximation in restless bandit problems, such as the works of [1,14],

© The Author(s), under exclusive license to Springer Nature Switzerland AG 2025
A. Devos et al. (Eds.): ASMTA 2024, LNCS 14826, pp. 67–80, 2025.
https://doi.org/10.1007/978-3-031-70753-7_5

and [11], which use deep reinforcement learning to approximate the Whittle indices used in their heuristics.

One can generalize the restless bandits to weakly coupled MDPs, where the independent MDPs are coupled only through a constraint on the action and actions can belong to complex spaces. These problems present substantial complexity due to constraints of the actions and common resources. A key advancement in addressing such complex problems came with the introduction of Lagrangian Decomposition methods, as explored by [7]. The approach of [7] proposes a Lagrangian decomposition approach for solving the weakly coupled dynamic optimization problem, which yields upper bounds as well as heuristic solutions. Works by [10,16] have introduced methods for navigating these complex decision spaces, employing Gaussian processes and simulation-based algorithms, respectively, to tackle the multi-action challenges.

Other studies in weakly coupled MDPs include the work of [20], which addresses the challenges of online learning in this specific MDP setting and presents an algorithm with a tight $O(\sqrt{t})$ regret and constraint violations simultaneously. Additionally, [5] introduces the LP-update policy, which generalizes the classical restless bandit problems and demonstrates asymptotic optimality at various rates depending on problem characteristics.

Significant advances in deep reinforcement learning include the development of Deep Deterministic Policy Gradient (DDPG) [9] and Twin Delayed DDPG (TD3) [4], algorithms that have significantly advanced complex control tasks by solving MDPs with continuous actions. Building on the capabilities of these frameworks, the OptLayer architecture was introduced [12], specifically designed to generate safe, constraint-compliant actions. OptLayer integrates an additional layer that solves a constraint optimization problem applicable to both DDPG and TD3 architectures. This extension ensures that the actions taken by the learning models adhere to predefined constraints. [8] explores the online learning landscape for discrete multi-action RMABs and presents a Q-learning Lagrange policy algorithm tailored for restless multi-armed bandits with multiple discrete actions. Similarly, [15] uses this Lagrangian decomposition to train separate subagents for each individual MDP problem, and a general network to combine these results, also in the context of discrete multi-action RMABs.

In this work, we introduce the Lagrange Policy for Continuous Actions (LPCA) algorithm, a reinforcement learning algorithm specifically designed for weakly coupled MDP problems with continuous action spaces. To the best of our knowledge, this is the first paper proposing an algorithm to solve weakly coupled MDPs with continuous actions. LPCA integrates a neural network-based framework to study weakly coupled MDP using the Lagrange relaxation introduced in [7] to decouple the projects of the MDP, being able to study their dynamics independently of one another and effectively balancing resource constraints and individual project decisions. Continuous actions allow for a more accurate representation of real-world scenarios, such as adjusting resource levels or control parameters, without the limitations of discretization. This flexibility enhances the algorithm's ability to optimize performance by better managing trade-offs between competing processes, ultimately leading to more robust and efficient policy learning.

2   Problem Formulation

In our approach to the weakly coupled MDPs with continuous actions, we consider an environment consisting of N projects, each characterized by its unique state, action, and the resulting reward. Specifically, the state of the system is given by $\mathbf{s} = (s_1, \ldots, s_N) \in \mathbf{S}$, where each project is represented as s_i, an element from the finite state space S_i, $i = 1, \ldots, N$. Correspondingly, the actions taken in each project are denoted as elements a_i belonging to the compact action space A_i, and the complete system action is denoted with bold font $\mathbf{a} = (a_1, \ldots, a_N) \in \mathbf{A}$. The rewards obtained from these actions are encapsulated as elements r_i in the reward vector $\mathbf{r}$. The cost associated with each action a_i is expressed as $c(a_i)$, and the cumulative cost for all actions is given by $C(\mathbf{a}) = \sum_i c(a_i)$.

The system dynamics are governed by a transition probability kernel $T : \mathbf{S} \times \mathbf{A} \times \mathbf{S} \rightarrow [0,1]$, which specifies the probabilities of transitioning to new states given particular state and action vector. Given the values of actions, T has a product form. A discount factor $\gamma \in (0,1)$ is used to balance immediate and future rewards.

The long-term discounted reward can be expressed through the Bellman value function $V(\mathbf{s})$, which is the expected sum of discounted rewards accumulated over time, starting from the state $\mathbf{s}$ and satisfying the Bellman dynamic programming equation:

$$V(\mathbf{s}) = \max_{\mathbf{a} \in \mathbf{A}, \, C(\mathbf{a})=B} \left[\sum_{i=1}^{N} r_i(s_i, a_i) + \gamma \mathbb{E}[V(\mathbf{s}') \mid \mathbf{s}, \mathbf{a}] \right]. \tag{1}$$

The complexity of the problem comes primarily from the constraint imposed on the actions, which are dictated by a common pool of resources. Specifically, each project must select a continuous action $a_i \in [0, a_i^{\max}]$ whose activation cost, represented by the total cost $C(\mathbf{a})$, directly consumes a predefined total pool of available resources B. This shared resource pool constraint means that actions across projects are inherently coupled, which significantly increases the complexity of the decision space as the number of projects increases. The exponential growth in decision space complexity due to this coupling underscores the challenge of resource allocation and emphasizes the need for efficient use of the shared resource pool [2].

To manage this complexity, we can relax the value function using a Lagrange multiplier λ. This transforms the original problem into a Lagrangian form:

$$J(\mathbf{s}, \lambda) = \max_{\mathbf{a} \in \mathbf{A}} \left[\sum_{i=1}^{N} r_i(s_i, a_i) + \lambda \left(B - \sum_{i=1}^{N} c(a_i) \right) + \gamma \mathbb{E}[J(\mathbf{s}', \lambda) \mid \mathbf{s}, \mathbf{a}] \right]. \tag{2}$$

Here, λ is the Lagrange multiplier associated with the resource constraint B. By adjusting λ, we effectively balance the immediate cost of actions against their long-term rewards, allowing for a decoupling of the projects' decisions. If we assume the additive structure of the value function with respect to the projects of the weakly coupled MDP, the Eq. (2) can be rewritten as:

$$J(\mathbf{s}, \lambda) = \frac{\lambda B}{1 - \gamma} + \sum_{i=1}^{N} \max_{a_i \in A_i} Q_i(s_i, a_i, \lambda), \tag{3}$$

where

$$Q_i(s_i, a_i, \lambda) = r_i(s_i, a_i) - \lambda c(a_i) + \gamma \sum_{s_i'} T(s_i, a_i, s_i') \max_{a_i' \in A_i} Q_i(s_i', a_i', \lambda). \tag{4}$$

In this decoupled framework, the Lagrange multiplier λ is instrumental in determining the optimal policy for each project. Under the budget constraint B, λ acts as a trade off parameter by introducing a penalty term $\lambda c(a_i)$ for the actions taken. A higher λ parameter places more emphasis on minimizing the cost (i.e., staying within the resource limit B), while a lower λ value shifts the focus towards maximizing rewards with less emphasis on the cost implementations. As λ rises, the preferred policy for each project will increasingly favor actions that offer the highest "value-to-cost" ratio. Thus, the function (3) is a measure of the total expected reward, adjusted for the cost of the actions taken under that policy. To balance the expected rewards with the cost of actions, we need to find λ^* such that

$$\lambda^*(\mathbf{s}) = \arg\min_{\lambda} J(\mathbf{s}, \lambda). \tag{5}$$

This term is defined as the best trade-off between maximizing rewards and minimizing the cost of actions. It is at this point that the policy aligns with the time-averaged resource constraints, ensuring that the actions selected are not only rewarding but also resource-efficient.

Then, in a continuous action framework, at each time step t we aim to solve the following Knapsack-like optimization problem:

$$\max_{\mathbf{a} \in \mathbf{A}} \sum_{i=1}^{N} Q_i(s_i(t), a_i, \lambda^*(s_i)) \qquad s.t. \qquad \sum_{i=1}^{N} c(a_i) = B. \tag{6}$$

In the LPCA algorithm, described in detail next, we interpolate the curve of the Q-values $Q(s, a, \lambda)$ as functions of the Lagrange multiplier λ through a neural network. This curve is a convex function with respect to λ [7], making the minimization of (3) a simple one-dimensional convex optimization problem once the neural network is trained. For the optimization (6) we explore two approaches as outlined in Sects. 3.1 and 3.2.

3   LPCA Algorithm

In numerous practical applications, the model parameters, particularly expected rewards and transition probabilities, are often unknown or inaccessible. To

---

Algorithm 1. LPCA Training Process

Require: Environment, N_{iter}, Update frequency N, Batch size M, Policy method
Ensure: Train LPCA Model, Update Policy Dictionary
 1: Initialize Q-value neural network, policy dictionary, experience memory
 2: **for** iteration $= 1$ **to** N_{iter} **do**
 3:     Select and execute action $\mathbf{a}$, store $(\mathbf{s}, \mathbf{a}, \mathbf{r}, \mathbf{s}'$, done$)$
 4:     **if** memory $\geq M$ **then**
 5:         Update Q-values with mini-batch of M (Algorithm 2)
 6:     **end if**
 7:     **if** iteration mod $N = 0$ **then**
 8:         Update policy with Differential Evolution or Greedy (Algorithm 3)
 9:     **end if**
10: **end for**

---

address this, traditional reinforcement learning methods have been employed to learn those parameters [17]. However, a significant challenge arises in environments where the projects of the MDP are coupled. In these cases, the complexity of solving the problem increases exponentially with the number of projects. To address this challenge, we introduce LPCA, a reinforcement learning algorithm that extends Q-learning by incorporating neural networks for approximating Q-values for constrained continuous actions. This section details the operation and implementation of LPCA.

The core methodology of the LPCA algorithm involves a two-timescale process centered around learning and optimization. Initially, LPCA focuses on training a neural network to accurately approximate the Q-values as defined in Eq. (4). This process involves learning the balance between immediate rewards, action costs, and future rewards based on the transition dynamics of the system. Once the neural network is effectively trained, in online fashion, for the current coupled state $\mathbf{s}$, LPCA computes the value function $J(\mathbf{s}, \lambda)$ as described in Eq. (3). The objective is to determine the optimal Lagrange multiplier λ^* that minimizes $J(\mathbf{s}, \lambda)$ as formulated in Eq. (5). Finally, LPCA addresses the optimization problem set out in Eq. (6) through two possible methods: a differential evolution optimizer (Algorithm 4) or a greedy optimizer (Algorithm 5).

The general training process of LPCA, as outlined in Algorithm 1, is a key aspect of our approach. The algorithm begins by utilizing a policy dictionary to interact with the environment. This dictionary is a mapping of states to actions, where each state corresponds to a unique action vector. During each interaction, an action is selected based on the current policy, and the environment responds accordingly. The response, including the state transition and reward information, is stored as a transition sample. Notably, each process of the weakly coupled MDP is treated individually, with the transition sample from each project recorded separately in a memory buffer. This memory serves as a repository for experiences, which are later used to update the neural network that approximates Q-values.

Algorithm 2. Update Q-values in LPCA Neural Network Model

1: **for** each random sample in memory **do**
2:    Extract s, a, r, s', $is\_terminal$ from sample {$is\_terminal$ indicates if s' is a terminal state}
3:    $Q \leftarrow$ Calculate target Q-values for s and a using a subset of λ values lambda_grid

4:    $V_{expected} \leftarrow$ Calculate expected value functions for s' using target network for each $\lambda \in$ lambda_grid
5:    **if** $is\_terminal$ **then**
6:        $Q_{target}(s, a, \lambda) \leftarrow r(s) - \lambda c(a)$
7:    **else**
8:        $Q_{target}(s, a, \lambda) \leftarrow r(s) - \lambda c(a) + \gamma \cdot V_{expected}$
9:    **end if**
10:    Perform a gradient descent step on $(Q_{target}(s, a, \lambda) - Q(s, a, \lambda))^2$ to update network weights
11: **end for**
12: Perform soft-update on target network weights $\theta' \leftarrow \theta\tau + (1 - \tau)\theta'$

The training of the neural network, as detailed in Algorithm 2, is central to learning the Q-values from Eq. (4) associated with state transitions (s, a, r, s') across a range of test λ values. These test values are selected as a random subset from 'lambda_grid', which encompasses a discretized set of λ values in the range of a problem-dependent $[-\lambda_{\max}, \lambda_{\max}]$, using 1000-point discretization.

During each iteration of the training process, the algorithm samples a batch of experiences from the memory. Each experience comprises the current state s, the action taken a, the reward received r, the subsequent state s', and a boolean flag indicating the terminal status of s', i.e. whether s' is the last state in an epoch, for a given individual project. For each experience, the algorithm computes the target Q-values for the state-action pair (s, a) using a random subset of λ values from 'lambda_grid'. This step involves evaluating the Q-value function for different levels of resource utilization and cost. By using a random subset of λ values, the algorithm optimizes computation, reducing the number of evaluations needed for each update. Additionally, this approach helps to avoid overfitting by selecting different λ points each time, ensuring that the model does not become too specialized to specific values of λ. The computation of the target Q-values $Q_{target}(s, a, \lambda)$ utilizes a target network, which is a lagged version of the primary neural network, to provide stable targets for learning [18].

Through this training process, the LPCA algorithm efficiently learns the Q-values for various state transitions under different levels of resource constraints, as dictated by the varying λ values.

Having trained the neural network to generate accurate approximations of Eq. (4), we proceed with Algorithm 3 to compute the value function $J(\mathbf{s}, \lambda)$ for a given state $\mathbf{s}$ as in Eq. (3). This computation involves evaluating $\sum_{i=1}^{N} \max_{a_i} Q(s_i, a_i, \lambda)$ for every λ within the discretized set 'lambda_grid'.

Algorithm 3. Computation of Lagrange term λ^*

Require: method
Ensure: Updated policy dictionary $\pi(\mathbf{s})$
 1: **function** PolicyDictUpdate(method)
 2: **for all** $\mathbf{s} \in \mathbf{S}$ **do**
 3:     q_table $\leftarrow$ Zero Matrix of size [n_lambda, N]
 4:     **for** $i \in 1 : N$ **do**
 5:         q_table$[:, i] \leftarrow \max_{a_i} Q(s_i, a_i, \lambda), \forall \lambda \in$ lambda_grid
 6:     **end for**
 7:     $J(\mathbf{s}, \lambda) \leftarrow$ Compute value functions as (3)
 8:     $\lambda^*(\mathbf{s}) \leftarrow \arg\min_\lambda J(\mathbf{s}, \lambda)$
 9:     **if** method = Evolution **then**
10:         $\mathbf{a}^* \leftarrow$ DifferentialEvolution($\mathbf{s}, \lambda^*(\mathbf{s}), a_{\max}$)
11:     **else**
12:         $\mathbf{a}^* \leftarrow$ Greedy($\mathbf{s}, \lambda^*(\mathbf{s}), a_{\max}, \delta$)
13:     **end if**
14:     $\pi(\mathbf{s}) \leftarrow \mathbf{a}^*$
15: **end for**
16: **end function**

Once this term is calculated, obtaining the optimal λ^* is a one-dimensional convex optimization problem, as shown in Eq. (5).

A key technical contribution of our work is how we explore the action space to solve the knapsack problem described in Eq. (6). This problem is challenging in neural networks due to the existence of many local minima, where traditional gradient optimization methods get stuck.

We propose two different strategies to explore this action space in order to make the best use of the available resources and select the best action based on our Q-value estimates. The first strategy, presented in Sect. 3.1, is an evolutionary algorithm (LPCA-DE). It uses mechanisms similar to natural selection to iteratively search for the optimal solution, effectively avoiding local minima by exploring a wider range of solutions.

The second strategy, presented in Sect. 3.2, is a greedy algorithm (LPCA-Greedy). It focuses on choosing the action based on the gradient of the Q-values with respect to the actions for each project, selecting the action that promises the highest increase in the Q-value per unit of resource expended. This method is simpler and faster, and helps to quickly identify actions that increase payoff, even if it does not explore as widely.

3.1 Differential Evolution Optimization (LPCA-DE)

The first method (Algorithm 4) employs a differential evolution algorithm, renowned for its effectiveness in identifying global optima and circumventing local optima traps. This method is particularly adept at exploring the search space comprehensively [3].

A critical aspect of this approach is the integration of a penalty mechanism to ensure that action selection remains within resource constraints. Actions leading to resource utilization beyond the available limit are subjected to a significant penalty. This mechanism is in line with the role of the λ term in the Q-value definition (see Eq. (4)). Given the $\lambda c(a)$ term in Eq. (4), the derived optimal policy tends towards cost-effectiveness. However, it may not always coincide with the optimal policy of the original constrained problem (see Eq. (1)) particularly if a higher action's benefit does not justify its cost in the relaxed problem, leading to potential underutilization of resources. This leads to a policy that may not fully utilize the available resources as defined in Eq. (6). To address this, we introduce an additional penalty, proportional to the amount of unused resources, into the differential evolution optimization problem. This modification guides the optimizer towards actions that maximize resource usage, ensuring the algorithm not only pursues cost-effective solutions but also fully utilizes the available resources.

Algorithm 4. Action Selection through Differential Evolution Optimization

---
Require: State vector $\mathbf{s}$, fixed Lagrange multiplier λ_{fix}, maximum action $a_{\max}$
Ensure: Optimal actions maximizing Q-values under resource constraints
1: **function** DifferentialEvolution($\mathbf{s}, \lambda_{\text{fix}}, a_{\max}$)
2:   Bounds $\leftarrow [0, a_{\max}]$
3:   **function** ObjectiveFunction($\mathbf{a}, \mathbf{s}, \lambda^*$)
4:     $Q_{\text{total}} \leftarrow \sum_{i=1}^{N} Q(s_i, a_i, \lambda^*)$
5:     $C_{\text{total}} \leftarrow \sum_{i=1}^{N} C(s_i, a_i)$
6:     **if** $C_{\text{total}} > B$ **then**
7:       Penalty $\leftarrow$ Large constant value
8:       $Q_{\text{total}} \leftarrow Q_{\text{total}} -$ Penalty
9:     **else if** $C_{\text{total}} < B$ **then**
10:      Penalty $\leftarrow B - C_{\text{total}}$
11:      $Q_{\text{total}} \leftarrow Q_{\text{total}} -$ Penalty
12:    **end if**
13:    **return** $-Q_{\text{total}}$
14: **end function**
15: $\mathbf{a}^* \leftarrow$ Apply Differential Evolution optimization with ($ObjectiveFunction, Bounds$)
16: **return** $\mathbf{a}^*$
17: **end function**

3.2   Greedy Optimization Strategy (LPCA-Greedy)

The second method (Algorithm 5) is a greedy optimization strategy. This approach is characterized by its iterative process of evaluating the gradient of the Q-values with respect to the actions for each project and then allocating resources to the project with the highest gradient. The process continues until all resources are exhausted.

This strategy prioritizes complete resource utilization, assigning resources to the projects that promise the highest increase in the Q-value per unit of resource expended. Unlike the differential evolution method, which searches for an optimal policy and then adjusts for resource utilization, the greedy approach begins with the premise of full resource allocation and does so in a manner that maximizes the benefit derived from each project.

The choice between these methods can be guided by the specific characteristics of the problem at hand, such as the nature of the resource constraints and the desired balance between resource utilization and reward maximization.

Algorithm 5. Greedy Action Selection for Continuous MDP
___

Require: State $\mathbf{s}$, λ_{fix}, max action $a_{\max}$, increment δ
Ensure: Optimal actions maximizing Q-values, maximum action $a_{\max}$
1: **function** Greedy($\mathbf{s}, \lambda_{\text{fix}}, a_{\max}, \delta$)
2: Initialize action vector $\mathbf{a}$ to zeros, $B_{\text{remaining}} = B$
3: **while** $B_{\text{remaining}} > 0$ **do**
4:     $i \leftarrow \arg\max_i \frac{\partial Q}{\partial a_i}$
5:     $a_i \leftarrow a_i + \delta$, ensure $a_i \le a_{\max}$
6:     $B_{\text{remaining}} \leftarrow B - \sum_{i=1}^{N} c(s_i, a_i)$
7: **end while**
8: **return a**
9: **end function**

---

4   Experimental Results

To evaluate the effectiveness of our algorithms, we rely on measuring the average discounted rewards that their policies yield. Given a discount factor of $\gamma = 0.9$, we examine the rewards that each algorithm's policy yields over $t \in [0, 50]$ iterations, starting from every possible state in our problem space. The evaluation process involves computing the discounted sum of the rewards using the equation

$$R = \sum_{t=0}^{50} \sum_{i=1}^{N} \gamma^t r(s_i(t), a_i(t)),$$

where $r(s_i(t), a_i(t))$ represents the reward received at time t for being in state $s_i(t)$ and taking action $a_i(t)$, for each MDP i. To ensure statistical robustness and to derive confidence intervals for our performance metrics, we repeat this evaluation 100 times. The results are shown in our figures, with the mean performance represented by bold lines and the confidence intervals represented by the shaded areas surrounding these lines.

Our experimental framework encompasses three distinct types of problems: Type A and Type B, each representing a continuous action version of challenges similar to those discussed in [8], and the speed scaling problem inspired from [22]. Types A and B feature two states per project with $a \in [0, 2]$, with a reward function $R(s) = s$ and a cost function $C(a) = a$. The key difference between Type A and Type B lies in their transition probability matrices:

$$P_A(a) = \begin{pmatrix} 0.02a^2 - 0.09a + 0.8 & -0.02a^2 + 0.09a + 0.2 \\ 0.75e^{-0.947a} & 1 - 0.75e^{-0.947a} \end{pmatrix}$$

$$P_B(a) = \begin{pmatrix} 0.95e^{-2.235a} & 1 - 0.95e^{-2.235a} \\ 0.3347e^{-1.609a} & 1 - 0.3347e^{-1.609a} \end{pmatrix}.$$

Additionally, we introduce a mixed environment where half of the projects follow the transition probabilities of Type A and the other half those of Type B.

The speed scaling environment involves projects with six states, and $a \in [0, 2]$. We apply the uniformization technique [13] to construct an equivalent discrete time version of the continuous time problem. The transition probabilities are given by

$$P(a) = \begin{pmatrix} 1 - \frac{\alpha}{\nu} & \frac{\alpha}{\nu} & 0 & 0 & 0 & 0 \\ \frac{\mu_a}{\nu} & 1 - \frac{\alpha + \mu_a}{\nu} & \frac{\alpha}{\nu} & 0 & 0 & 0 \\ 0 & \frac{\mu_a}{\nu} & 1 - \frac{\alpha + \mu_a}{\nu} & \frac{\alpha}{\nu} & 0 & 0 \\ 0 & 0 & \frac{\mu_a}{\nu} & 1 - \frac{\alpha + \mu_a}{\nu} & \frac{\alpha}{\nu} & 0 \\ 0 & 0 & 0 & \frac{\mu_a}{\nu} & 1 - \frac{\alpha + \mu_a}{\nu} & \frac{\alpha}{\nu} \\ 0 & 0 & 0 & 0 & \frac{\mu_a}{\nu} & 1 - \frac{\mu_a}{\nu} \end{pmatrix},$$

where $\alpha = 0.9$ is the arrival rate, $\mu(a) = \sqrt{a}$ is the controlled departure rate, $\nu = \max_a(\alpha + \mu(a))$ is the normalization factor, β is the continuous discount factor related to the discrete factor γ as $\beta = \frac{\nu}{\gamma} - \nu$. The reward function is defined as:

$$R(s) = \frac{-s}{\nu + \beta} + \frac{C_r}{\nu + \beta} = \begin{cases} \frac{-s}{\nu + \beta} & \text{if } s < s_{\max} \\ \frac{-s_{\max} - 10}{\nu + \beta} & \text{if } s = s_{\max} \end{cases}$$

where $C_r = -10$ is the rejection cost that occurs in the final state $s_{\max}$. The cost function is defined as $C(s, a) = \frac{a}{\nu + \beta}$ if $s > 0$, otherwise 0.

For Types A and B, we conducted experiments with both 4 projects with 2 units of resources and 6 projects with 4 units of resources. The mixed environment, combining Types A and B, was tested with 6 projects and 4 units of

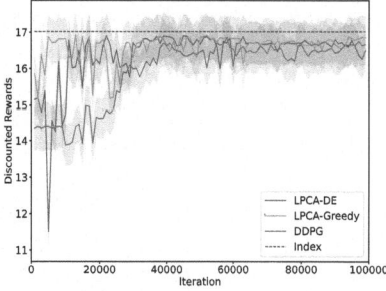 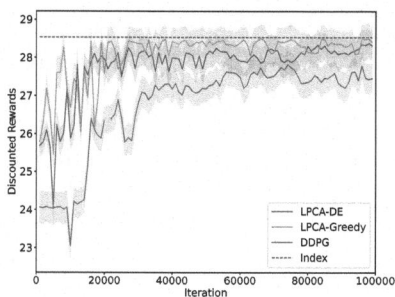

Fig. 1. Experimental results for Type A environment: (Left) 4 projects and 2 units of resources, (Right) 6 projects and 4 units of resources.

resources. The Speed Scaling experiment involved 4 projects with 1.5 units of resources, equivalent to fully activating two of the four projects.

To benchmark our algorithm, we choose DDPG (Deep Deterministic Policy Gradient) [9] augmented with OptLayer [12] as the baseline. OptLayer enhances DDPG by incorporating a constraint optimization layer in the actor network, enabling the generation of actions that respect the constraints outlined in the original problem formulation (Eq. (1) and (6)).

In addition to this, we have benchmarked Whittle's index heuristic for continuous actions. These indices are computed through the algorithm proposed by [19] for discrete multi-action ($a \in [0, 1, 2, \dots]$) and adapted for an arbitrary discretization δ_a of the action ($a \in [0, \delta_a, 2\delta_a, \dots]$). For an approximation of a fully continuous action, we use a discretization of $\delta_a = 0.001$, leading to a total of 2001 possible actions. Due to the large amount of indices to compute, a tabular learning algorithm for those indices would not be feasible.

In the 4 projects and 2 resources configuration, both LPCA-DE and LPCA-Greedy demonstrated a clear advantage over DDPG, particularly in Type B environment (Fig. 2 left), where the gap between the performance of both versions of LPCA and DDPG is larger and both LPCA algorithms converge to the Whittle Index policy performance. In Fig. 1 left, although DDPG achieves

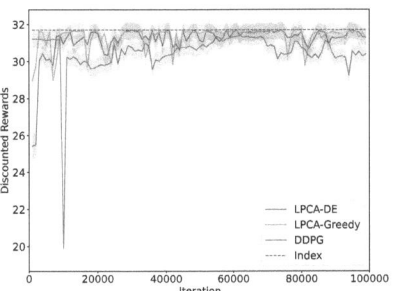
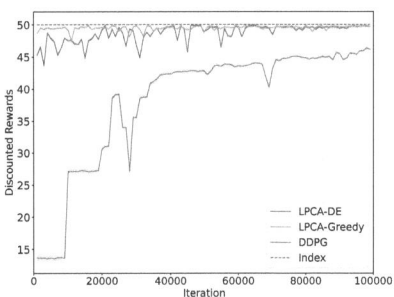

Fig. 2. Experimental results for Type B environment: (Left) 4 projects and 2 units of resources, (Right) 6 projects and 4 units of resources.

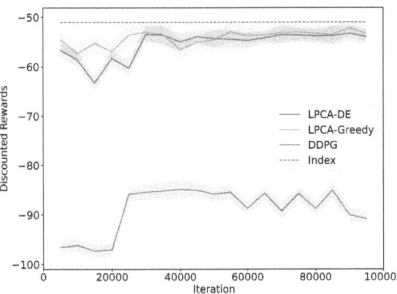
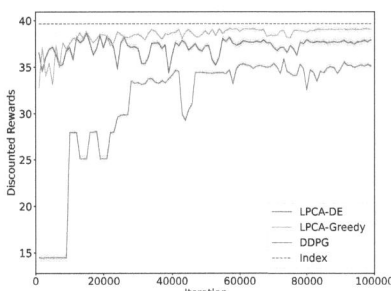

Fig. 3. (Left) Speed Scaling with 4 projects and 1.5 units of resources, (Right) Mixed Type A and B environments with 6 projects and 4 units of resources.

a similar level of performance to LPCA, the latter converges to a performance level similar to Whittle indices' much faster, while DDPG takes around 40000 iterations.

This gap widened significantly in the 6 projects and 4 resources setting. In Type A (Fig. 1 right), the optimality gap between both versions of LPCA and DDPG widens. A similar pattern shows in Type B (Fig. 2 right), with DDPG having subpar performance. In the mixed environment (Fig. 3 right), DDPG's performance reflects the issues observed in the previous scenarios. On the other hand, LPCA-DE and specially LPCA-Greedy are able to obtain a better policy, close to the Whittle index policy performance.

In the Speed Scaling experiment (Fig. 3 left), the performance of both LPCA-DE and LPCA-Greedy algorithms converges to a similar performance to the Whittle index policy, while DDPG's performance lags behind.

Overall, the LPCA algorithms consistently outperformed DDPG with Opt-Layer across various settings and environments. Notably, LPCA's superiority became increasingly pronounced in more complex scenarios involving a greater number of processes and limited resources.

5   Conclusion

In this study, we introduced the LPCA (Lagrange Policy for Continuous Actions) algorithm, a reinforcement learning approach for weakly coupled MDPs with continuous actions and resource constraints. Our experimental results demonstrate that LPCA, in both its Differential Evolution (DE) and Greedy variants, consistently outperforms the DDPG algorithm augmented with OptLayer across various scenarios. Notably, LPCA exhibits superior scalability with an increasing number of projects.

As a direction for future research, we aim to test the LPCA algorithm in larger-scale environments featuring more states per projects. This expansion will allow us to further evaluate LPCA's scalability and effectiveness in even more complex and dynamic settings, potentially broadening its applicability to a wider array of practical problems in operations research and beyond. The exploration of LPCA's performance in these extended scenarios is expected to yield valuable insights into its capabilities and limitations, guiding future enhancements and adaptations of the algorithm.

Acknowledgments. F. Robledo has received funding from the Department of Education of the Basque Government through the Consolidated Research Group MATH-MODE (IT1456-22). Research partially supported by the French "Agence Nationale de la Recherche (ANR)" through the project ANR-22-CE25-0013-02 (ANR EPLER) and DST-Inria Cefipra project LION.

References

1. Avrachenkov, K.E., Borkar, V.S.: Whittle index based Q-learning for restless bandits with average reward. Automatica **139**, 110186 (2022). https://doi.org/10.1016/j.automatica.2022.110186
2. Bertsekas, D.: Dynamic Programming and Optimal Control: Volume I. Athena Scientific (2012), google-Books-ID: qVBEEAAAQBAJ
3. Das, S., Suganthan, P.N.: Differential evolution: a survey of the state-of-the-art. IEEE Trans. Evol. Comput. **15**(1), 4–31 (2011). https://doi.org/10.1109/TEVC.2010.2059031
4. Fujimoto, S., Hoof, H., Meger, D.: Addressing Function Approximation Error in Actor-Critic Methods. pp. 1587–1596. PMLR, July 2018. https://proceedings.mlr.press/v80/fujimoto18a.html
5. Gast, N., Gaujal, B., Yan, C.: The LP-update policy for weakly coupled Markov decision processes. Tech. rep., November 2022. https://doi.org/10.48550/arXiv.2211.01961, http://arxiv.org/abs/2211.01961, arXiv:2211.01961 [math] type: article
6. Gittins, J.C.: Bandit processes and dynamic allocation indices. J. Roy. Stat. Soc.: Ser. B (Methodol.) **41**(2), 148–164 (1979). https://doi.org/10.1111/j.2517-6161.1979.tb01068.x
7. Hawkins, J.T.: A Langrangian decomposition approach to weakly coupled dynamic optimization problems and its applications. Ph.D. thesis, Massachusetts Institute of Technology (2003)
8. Killian, J.A., Biswas, A., Shah, S., Tambe, M.: Q-learning lagrange policies for multi-action restless bandits. In: Proceedings of the 27th ACM SIGKDD Conference on Knowledge Discovery & Data Mining, KDD 2021, pp. 871–881. Association for Computing Machinery, New York, August 2021. https://doi.org/10.1145/3447548.3467370
9. Lillicrap, T.P., Hunt, J.J., Pritzel, A., Heess, N., Erez, T., Tassa, Y., Silver, D., Wierstra, D.: Continuous control with deep reinforcement learning. Tech. rep., July 2019. https://doi.org/10.48550/arXiv.1509.02971, http://arxiv.org/abs/1509.02971, arXiv:1509.02971 [cs, stat] type: article
10. Meshram, R., Kaza, K.: Simulation Based Algorithms for Markov Decision Processes and Multi-Action Restless Bandits. Tech. rep., July 2020. https://doi.org/10.48550/arXiv.2007.12933, http://arxiv.org/abs/2007.12933, arXiv:2007.12933 [cs, eess] type: article
11. Nakhleh, K., Ganji, S., Hsieh, P.C., Hou, I.H., Shakkottai, S.: NeurWIN: neural whittle index network for restless bandits via deep RL. In: Advances in Neural Information Processing Systems, vol. 34, pp. 828–839. Curran Associates, Inc. (2021). https://proceedings.neurips.cc/paper/2021/hash/0768281a05da9f27df178b5c39a51263-Abstract.html
12. Pham, T.H., De Magistris, G., Tachibana, R.: OptLayer - practical constrained optimization for deep reinforcement learning in the real world. In: 2018 IEEE International Conference on Robotics and Automation (ICRA), pp. 6236–6243, May 2018. https://doi.org/10.1109/ICRA.2018.8460547, iSSN: 2577-087X
13. Puterman, M.L.: Markov Decision Processes: Discrete Stochastic Dynamic Programming. John Wiley & Sons, August 2014, google-Books-ID: VvBjBAAAQBAJ
14. Robledo, F., Borkar, V.S., Ayesta, U., Avrachenkov, K.: Tabular and deep learning of whittle index. In: EWRL 2022 - 15th European Workshop of Reinforcement Learning. Milan, Italy, September 2022. https://hal.science/hal-03767324

15. Shar, I.E., Jiang, D.: Weakly coupled deep Q-Networks. In: Oh, A., Neumann, T., Globerson, A., Saenko, K., Hardt, M., Levine, S. (eds.) Advances in Neural Information Processing Systems, vol. 36, pp. 43931–43950. Curran Associates, Inc. (2023)

16. Srinivas, N., Krause, A., Kakade, S.M., Seeger, M.: Gaussian process optimization in the bandit setting: no regret and experimental design. IEEE Trans. Inf. Theory **58**(5), 3250–3265 (2012). https://doi.org/10.1109/TIT.2011.2182033, http://arxiv.org/abs/0912.3995, arXiv:0912.3995 [cs]

17. Sutton, R.S., Barto, A.G.: Reinforcement learning: An introduction. MIT press (2018)

18. Van Hasselt, H., Guez, A., Silver, D.: Deep reinforcement learning with double q-learning. In: Proceedings of the AAAI Conference on Artificial Intelligence, vol. 30 (2016). https://ojs.aaai.org/index.php/AAAI/article/view/10295

19. Weber, R.: Comments on: dynamic priority allocation via restless bandit marginal productivity indices. TOP **15**(2), 211–216 (2007)

20. Wei, X., Yu, H., Neely, M.J.: Online learning in weakly coupled markov decision processes: a convergence time study. In: Proceedings of the ACM on Measurement and Analysis of Computing Systems 2(1), 12:1–12:38, April 2018. https://doi.org/10.1145/3179415

21. Whittle, P.: Restless bandits: activity allocation in a changing world. J. Appl. Probability **25**(A), 287–298 (1988). https://doi.org/10.2307/3214163.   https://www.cambridge.org/core/journals/journal-of-applied-probability/article/abs/restless-bandits-activity-allocation-in-a-changing-world/DDEB5E22AFFEFF50AA97ADC96B71AE35

22. Wierman, A., Andrew, L.L.H., Tang, A.: Power-aware speed scaling in processor sharing systems. In: IEEE INFOCOM 2009. pp. 2007–2015, April 2009. https://doi.org/10.1109/INFCOM.2009.5062123, https://ieeexplore.ieee.org/abstract/document/5062123, iSSN: 0743-166X

A Lazy Abstraction Algorithm
for Markov Decision Processes
Theory and Initial Evaluation

Dániel Szekeres[(✉)] [iD], Kristóf Marussy[iD], and István Majzik[iD]

Department of Measurement and Information Systems, Budapest University of
Technology and Economics, Budapest, Hungary
{szekeres,marussy,majzik}@mit.bme.hu

Abstract. Analysis of Markov Decision Processes (MDP) is often hindered by state space explosion. Abstraction is a well-established technique in model checking to mitigate this issue. This paper presents a novel lazy abstraction method for MDP analysis based on adaptive simulation graphs. Refinement is performed only when new parts of the state space are explored, which makes partial exploration techniques like Bounded Real-Time Dynamic Programming (BRTDP) retain more merged states. Therefore, we propose a combination of lazy abstraction and BRTDP. To evaluate the performance of our algorithm, we conduct initial experiments using the Quantitative Verification Benchmark Set.

Keywords: Lazy abstraction · Markov Decision Processes ·
Abstraction refinement · Probabilistic model checking

1 Introduction

Ensuring the reliable operation of safety-critical systems, like railway interlocking systems and embedded controllers, is vital. Probabilistic model checking addresses this by offering an automated approach with formal mathematical guarantees for the analysis of quantitative properties, like reliability and availability [20]. We focus on a fundamental task in probabilistic model checking: computing the worst-case probability of reaching an error state.

Markov Decision Processes (MDPs) are discrete-time models able to describe both probabilistic and non-deterministic behavior, used in reliability and safety analysis for worst-case modeling of unknown factors. The analysis of other modeling formalisms, like Markov Automata or Probabilistic Timed Automata, can often be reduced to MDP analysis as well.

The project supported by the Doctoral Excellence Fellowship Programme (DCEP) is funded by the National Research Development and Innovation Fund of the Ministry of Culture and Innovation and the Budapest University of Technology and Economics, under a grant agreement with the National Research, Development and Innovation Office.

A. Devos et al. (Eds.): ASMTA 2024, LNCS 14826, pp. 81–96, 2025.
https://doi.org/10.1007/978-3-031-70753-7_6

State space explosion presents an obstacle for MDP model checking: as the number of components or variables increases, the state space may grow exponentially. Consequently, practical implementations face problems in representing the system in memory and the numerical solution methods also become intractable.

Abstraction aims to counteract this. Several abstraction-based techniques have been adapted to probabilistic systems, like CEGAR [15,17] and abstract interpretation [7]. Partial state space exploration, like *Bounded Real-Time Dynamic Programming (BRTDP)* [3,18] is another approach for counteracting it. As most existing MDP abstraction methods rely on computing the whole abstract model to choose a refinement, they do not lend themselves well to combination with partial state space exploration techniques.

Lazy abstraction [13,23] in contrast merges state-space exploration and refinement, making it a good candidate for this combination. However, no such method has been proposed for MDPs to our knowledge.

We adapt an existing lazy abstraction algorithm [27] to MDPs (Sect. 3). We combine it with BRTDP, benefiting from the synergy of lazy abstraction and partial state space exploration and enabling a trade-off between time and accuracy (Subsect. 3.1). We evaluate the performance of the proposed algorithms using models from the Quantitative Verification Benchmark Set [12] (Sect. 4).

Related Work. *Counterexample-Guided Abstraction Refinement (CEGAR)* [6] is a successful approach for abstraction-based model checking: it starts with a coarse abstraction and refines it based on abstract counterexamples.

Lazy abstraction, introduced in [13], improved CEGAR through *on-demand refinement* during abstract state space exploration and varying precision from node to node in the state graph. An *interpolant-based version* was proposed in [23]. This was adapted to timed automata [14], introducing *Adaptive Simulation Graphs (ASG)* as the abstract model. This allows *earlier refinement*, cutting spurious paths before reaching a target, and a less expensive covering check. The ASG-based algorithm was adapted to explicit value abstraction of *discrete variables* in [27], which we, in turn, adapt to MDPs.

Different abstraction methods have been proposed for probabilistic systems. While some CEGAR-based methods employ MDPs as abstraction [5,8,15], others utilize stochastic games [17,24,28], which we plan to incorporate in the future. Abstract interpretation has also been used for probabilistic systems [7,11]. Some others include magnifying lens abstraction [9], which explores the whole concrete state space but keeps only its subset in memory and assume-guarantee-style abstraction [19], specialized for composite systems. To our knowledge, no lazy abstraction method has been proposed for probabilistic models yet.

The algorithm presented in this paper uses a symmetric representation constraint, resulting in an approach similar to *bisimulation reduction* techniques [10,16]. The main difference is that until the whole ASG is explored, only a limited version of "bisimilarity" holds which does not take the unexplored part of the state space into account, allowing coarser partitions on the already explored part. When combined with partial exploration, the algorithm can stop before exploring the full ASG. *Finite-horizon bisimulation minimization* [16] and *incremental bisimulation abstraction-refinement* [25] are similar in that they employ a relaxed version of bisimulation. Both of them limit the bisimilarity to a fixed path length and compute exact quotients w.r.t the relaxed relation, while we base the relaxation on the currently explored state space and do not aim for computing the coarsest relation.

BRTDP was introduced in [22] for Stochastic Shortest Paths, and [3] applied it to *general MDPs*. We combine it with our lazy abstraction algorithm.

2   Background and Notations

$\mathbb{D}(A)$ is the set of probability distributions over the set A. For $d \in \mathbb{D}(A), a \in A$, $d(a)$ denotes the probability measure of a according to d. $f \colon A \hookrightarrow B$ means f is a *partial function* from A to B, and $Supp(f)$ is the set of values for which f is defined. For $d \in \mathbb{D}(A)$, $Supp(d) = \{a \in A | d(a) > 0\}$. δ_x is a Dirac distribution: $\delta_x(x) = 1, \forall y \neq x \colon \delta_x(y) = 0$.

2.1   Markov Decision Process (MDP)

MDPs are low-level mathematical models that describe both probabilistic and non-deterministic behavior in discrete time.

Definition 1 (MDP). *An MDP is a tuple $M = (S, Act, T, s_0)$, where S is the set of states, Act is the set of actions, $T \colon S \times Act \times S \to [0, 1]$ is a probabilistic transition function s.t. $\forall s \in S, a \in Act \colon \sum_{s' \in S} T(s, a, s') \in \{0, 1\}$ and $s_0 \in S$ is the initial state.*

An action $a \in Act$ is *enabled* in $s \in S$ if $\sum_{s' \in S} T(s, a, s') = 1$. In this case, $T(s, a) \in \mathbb{D}(S)$ denotes the next state distribution after taking a in s, defined as $T(s, a)(s') = T(s, a, s')$. The intuitive behavior of an MDP is as follows: starting in s_0 an action a is chosen non-deterministically from those enabled in the current state s_i in each step, and the next state is sampled from $T(s_i, a)$. A *trace* of an MDP is an alternating list of states and actions $s_0 \xrightarrow{a_1} s_1 \xrightarrow{a_2} s_2 \xrightarrow{a_3} \ldots$ such that $\forall i \colon T(s_{i-1}, a_i, s_i) > 0$. Fixing a strategy for resolving the non-determinism, the set of traces can be equipped with a probability measure: intuitively, the probability of a trace is the product of the probability of landing in each state of the trace after taking the action specified by the strategy in the previous state. For a detailed formal treatment, see e.g. [20].

Given an MDP of the system behavior and a set of target (error) states E, we want to compute (an upper approximation of) the probability of the set of traces involving a state in E with non-determinism resolved by a maximizing strategy: $\mathbb{P}_{max}(\{(s_0 \xrightarrow{a_1} s_1 \xrightarrow{a_2} s_2 \xrightarrow{a_3} \ldots) \mid \exists i \in \mathbb{Z}^+ \colon s_i \in E\})$. The result is the same if we make all target states absorbing, allowing us to restrict the analysis to finite traces.

Symbolic MDPs. Most real-life models are specified symbolically using state variables and operations on them. We assume that the MDP is given by a set of variables $\mathcal{V}$ and a set of probabilistic guarded commands $\mathcal{C}$. Each $v \in \mathcal{V}$ has a set $R(v)$ of values it can take, and an initial value $v_0 \in R(v)$. A *valuation* over $\mathcal{V}$ is a function $val: \mathcal{V} \to \bigcup_{v \in \mathcal{V}} R(v)$ s.t. $\forall v \in \mathcal{V}: val(v) \in R(v)$, and $VAL_\mathcal{V}$ is the set of all valuations over $\mathcal{V}$. The initial valuation of the model is a valuation val_0 s.t. $\forall v \in \mathcal{V}: val_0(v) = v_0$. The state space of this MDP is a subset of $VAL_\mathcal{V}$, and its initial state is val_0.

Let $\mathcal{B}_\mathcal{V}$ denote the set of Boolean expressions over $\mathcal{V}$, $\mathcal{E}_\mathcal{V}^v$ the set of expressions over $\mathcal{V}$ that result in an element of $R(v)$, and $\mathcal{E}_\mathcal{V} = \bigcup_{v \in \mathcal{V}} \mathcal{E}_\mathcal{V}^v$. An *assignment* is a function $a: \mathcal{V} \to \mathcal{E}_\mathcal{V}$, such that $\forall v \in \mathcal{V}: a(v) \in \mathcal{E}_\mathcal{V}^v$. Let $\mathcal{A}_\mathcal{V}$ denote the set of assignments for $\mathcal{V}$. $eval(e, val)$ for $e \in \mathcal{E}_\mathcal{V}$ and $val \in VAL_\mathcal{V}$ is the constant resulting from replacing each $v \in \mathcal{V}$ in e with $val(v)$. $eval(a, val)$ for $a \in \mathcal{A}_\mathcal{V}$ is a valuation val' such that $\forall v \in \mathcal{V}: val'(v) = eval(a(v), val)$. $eval(d, val)$ for $d \in \mathbb{D}(\mathcal{A}_\mathcal{V})$ is the distribution $d' \in \mathbb{D}(VAL_\mathcal{V})$ such that $d'(val') = \sum_{a \in \{a \in Supp(d) \ | \ eval(a, val) = val'\}} d(a)$.

A *command* $c \in \mathcal{C}$ consists of a guard $g_c \in \mathcal{B}_\mathcal{V}$ and a result distribution over assignments $d_c \in \mathbb{D}(\mathcal{A}_\mathcal{V})$. c is *enabled by val* iff $eval(g_c, val) = True$. Let $a_i^c \in Supp(d_c)$ denote the ith assignment of command c for a fixed ordering. The enabled actions in each state val of the represented MDP are the commands enabled by val, and taking the command c results in the distribution $eval(d_c, val)$. Widely used MDP description formats, like that of PRISM [21] or the JANI [4] format can be mapped to this low-level description.

Our running example is given by the following variables and commands:

$$\mathcal{V} = \{x, y\}, R(x) = R(y) = \mathbb{N}, x_0 = y_0 = 0$$
$$\mathbf{c_1}: [true] \ 0.8 : (x' := x + 1 \land y' := y), 0.2 : (x' := x \land y' := y)$$
$$\mathbf{c_2}: [x == 0] \ 1.0 : (x' := 1 \land y' := 2)$$
$$\mathbf{c_3}: [x == 2 \land y == 2] \ 1.0 : (x' := x \land y' := 3)$$

c_1 is enabled in every state, and it increments x by 1 with probability 0.8. c_2 is enabled when $x = 0$, and always sets y to 2 and x to 1. c_3 is enabled when x is 2 and y is 2, and sets y to 3. Figure 1a shows this MDP.

2.2 Lazy Abstraction

Abstraction-refinement methods mitigate state-space explosion by disregarding information present in the original *concrete* model to create an *abstract* model that is iteratively refined until a conclusion is reached. Lazy abstraction performs refinement on-the-fly and only on a subset of the state space.

For checking safety properties in the qualitative case, a conservative abstraction overapproximates the *reachable state set*. In the probabilistic setting, the *probability* of reaching a target state in the abstract model overapproximates that in the concrete one, which we will prove for the proposed algorithm.

We build on the lazy abstraction method of [27] for non-probabilistic systems. It constructs an *Adaptive Simulation Graph (ASG)* with nodes labeled by both a concrete and an abstract state: the concrete state represents all states in the abstract state regarding possible action sequences. The abstract state labels start very coarse and are refined as needed. *Covering edges* indicate that action sequences starting from the coverer node encompass those starting from the covered node, eliminating the need to explore paths from the covered node.

If an action is enabled in at least one concrete state described by the abstract label of a node, but not in the concrete label, the abstract label is *strengthened* by removing states with the action enabled. This operation can trigger additional strengthenings. The algorithm terminates once all enabled actions in non-covered nodes have been explored. The abstract labels in the finished ASG cover all reachable concrete states, and contain a target state only if one is reachable.

Abstract Domains. Abstract states are described using an *abstract domain*. For a set of concrete states S, an abstract domain $D = (\hat{S}, \preceq, \alpha, \gamma)$ consists of the abstract state set $\hat{S}$, a partial ordering $\preceq \subseteq \hat{S} \times \hat{S}$, an abstraction function $\alpha \colon 2^S \to \hat{S}$ and a concretization function $\gamma \colon \hat{S} \to 2^S$ satisfying $\forall A \in 2^S, \hat{a} \in \hat{S} \colon \alpha(A) \preceq \hat{a} \iff A \subseteq \gamma(\hat{a})$. γ lets us treat abstract states as sets of concrete states; we write "$s \in \hat{s}$" for $s \in \gamma(\hat{s})$ when γ is clear from the context. $x \preceq y$ denotes $(x, y) \in \preceq$. $\hat{S}$ has two special elements: $\top$ and $\bot$ satisfying $\gamma(\top) = S, \gamma(\bot) = \{\}$.

Our lazy abstraction algorithms are domain agnostic, but need an abstract domain for $S = VAL_\mathcal{V}$ with the following operations.

For $a \in \mathcal{A}_\mathcal{V}$ and $\hat{s} \in \hat{S}$, *abstract post operator* $eval(a, \hat{s}) \in \hat{S}$ applies an assignment in the abstract state space: $eval(a, \hat{s}) = \alpha(\{eval(a, s) | s \in \hat{s}\})$. For $b \in \mathcal{B}_\mathcal{V}$, $eval(b, \hat{s}) \in \{True, False, Unknown\}$ denotes evaluating b in the abstract state space: *True* if b evaluates to *True* for *all* $s \in \hat{s}$, *False* if b evaluates to *False* for *all* $s \in \hat{s}$, otherwise *Unknown*.

We also need a *block* operation: for an abstract state $\hat{s} \in \hat{S}$, a Boolean expression $b \in \mathcal{B}_\mathcal{V}$ and a concrete state $s \in \hat{s}$ s.t. $eval(b, s) = False$, $\hat{s}' = block(\hat{s}, b, s)$ is an abstract state s.t. $\hat{s}' \preceq \hat{s}, s \in \hat{s}', eval(b, \hat{s}') = False$. Its goal is to give a new abstract state by removing at least those states from $\hat{s}$ which satisfy b (potentially others as well) while keeping s.

The abstract states must be representable as Boolean expressions: for each $\hat{s} \in \hat{S}$ a $b_{\hat{s}} \in \mathcal{B}_\mathcal{V}$ must exist s.t. $\forall s \in S \colon eval(b_{\hat{s}}, s) = True \iff s \in \hat{s}$. Relying on this, we will freely use abstract states in place of Boolean expressions.

We will use the *explicit value domain* D_{expl} (abstract states correspond to tracking only a subset of $\mathcal{V}$) as an example throughout the paper which we implemented in our prototype, along with predicate abstraction D_{pred} (abstract states are Boolean predicates over $\mathcal{V}$). A partial function $pval \colon \mathcal{V} \hookrightarrow \bigcup_{v \in \mathcal{V}} R(v)$ s.t. $\forall v \in Supp(pval) \colon pval(v) \in R(v)$ is called a *partial valuation*, $PVAL_\mathcal{V}$ denotes the set of all partial valuations over $\mathcal{V}$. Description of these domains and their operations can be seen in Table 1, assuming a concrete state set $VAL_\mathcal{V}$.

The lazy abstraction algorithm does not use the abstraction and concretization functions α and γ and the abstract post operator $eval(a, \hat{s})$ directly, only as arguments of a block operation (see later), so they need not be efficiently computable if the corresponding block operation can be implemented efficiently.

Table 1. Properties and operations of the explicit value and predicate domains

	EXPL	PRED
$\hat{S}$	$PVAL_{\mathcal{V}} \cup \{\bot_{expl}\}$	$\mathcal{B}_{\mathcal{V}}$
$\preceq$	$pval \preceq pval' \iff (Supp(pval') \subseteq Supp(pval) \wedge \forall v \in Supp(pval') : pval(v) = pval'(v))$	$b_1 \preceq b_2 \iff (b_1 \implies b_2)$
α	$Supp(\alpha(A)) = \{v \in \mathcal{V} \mid \exists k \in R(v) : \forall val \in A : val(v) = k\}$ and $\forall v \in Supp(\alpha(A)) : \alpha(A)(v) = k \iff \forall val \in A : val(v) = k$	$\bigvee_{val \in A}(\forall v \in \mathcal{V} : v = val(v))$
γ	$\gamma(pval) = \{val \in VAL_{\mathcal{V}} \mid \forall v \in Supp(pval) : val(v) = pval(v)\}$	$\gamma(b) = \{val \in VAL_{\mathcal{V}} \mid val \vdash b)\}$
$\top, \bot$	$\top$ is the empty valuation, $\bot$ is a non-valuation element $\bot_{expl}$ representing contradiction	$\top = True, \bot = False$
Boolean representation	$b_{pval} = (\bigwedge_{v \in Supp(pval)} v = pval(v))$	identity (already a Boolean expression)
$eval(b, \hat{s})$	Substituting the values in $Supp(\hat{s})$, and deciding the satisfiability of the result.	$True$ if $\hat{s} \implies b$, $False$ if $\hat{s} \implies \neg b$, $Unknown$ if neither.
$eval(a, \hat{s})$	Substituting known variables into the result expressions. If this results in a constant, that is the result, else unknown.	strongest postcondition

3 Lazy Abstraction for MDPs

Now we adapt the lazy algorithm to symbolic MDPs given by a variable set $\mathcal{V}$ and a command set $\mathcal{C}_0$. Given a target formula $\phi \in \mathcal{E}_{\mathcal{V}}$, the goal is to compute the maximal probability of reaching a state s s.t. $eval(\phi, s) = True$.

We select abstract domain $(\hat{S}, \preceq, \alpha, \gamma)$. The set of commands is extended with a target command: $\mathcal{C} = \mathcal{C}_0 \cup \{(\phi, \delta_{id})\}$, where id is the identity assignment. A node is a target if this command is enabled in it. This ensures that the finished ASG contains a node labeled with a target state exactly if a target state is reachable in the concrete state space [27].

Abstract Model. We use a probabilistic extension of the ASG. A direct adaptation of the non-probabilistic lazy algorithm by switching to probabilistic actions would overapproximate the target probability with no control over the approximation. Therefore, we use a stricter, symmetric representation constraint for the relation between the concrete and abstract labels of a Probabilistic ASG node.

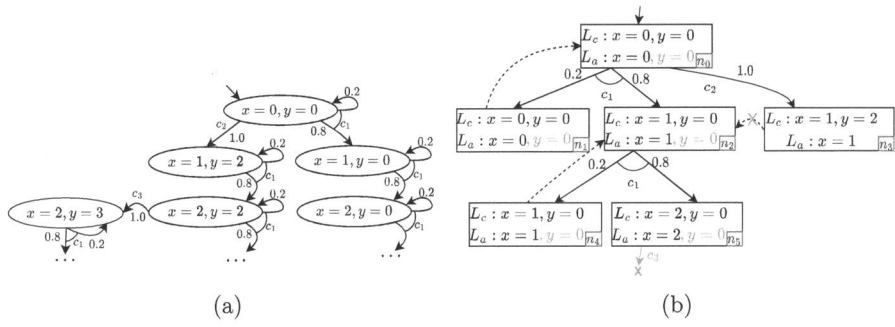

(a)   (b)

Fig. 1. Our running example MDP (a) and an in-progress PASG for it (b). The result of a refinement step is marked with orange

Definition 2 (*PASG*). *A Probabilistic Adaptive Simulation Graph is a tuple* (N, E_T, E_C, L_c, L_a), *where* N *is a set of nodes,* $E_T \subseteq N \times \mathcal{C} \times \mathbb{D}(N)$ *is a set of transition "edges" from nodes to node distributions labeled with commands,* $E_C \subseteq N \times N$ *is a set of directed covering edges,* $L_c : N \to VAL_{\mathcal{V}}$ *is the concrete labeling function,* $L_a : N \to \hat{S}$ *is the* abstract labeling function.

For an edge $e = (n, c, d) \in E_T$, n_i^e is the ith element of $Supp(d)$ for a fixed ordering. A PASG is *well-labeled*, if it satisfies the constraints in Table 2. The main difference from the original ASG is that in B2), we use a distribution of results instead of a single one, and A2) requires the set of enabled actions to be exactly the same in all concrete states contained in the abstract label.

In the original lazy algorithm, refinement is performed when an action *disabled* in $L_c(n)$, but *enabled* in some element of the abstract label $L_a(n)$, and we refine by blocking the guard of the action from $L_a(n)$. Here, we *also* refine when an action is *enabled* in $L_c(n)$ but *disabled* in some element of $L_a(n)$ by blocking the *negation* of the guard. We also need to adapt soundness to probabilities, see Theorem 1. D1) is a technical constraint to make our proofs easier.

Example 1. Figure 1b shows an example PASG with black (the orange part is a refinement example explained later). The abstract label tracks only x in all nodes (this could differ from node to node in general). L_c is contained in L_a for all nodes as the value of x is the same in L_c and L_a (A1)).

n_0 covers n_1, n_2 covers n_4 and n_3, satisfying C1) and C2). n_3 could cover n_4 according to the labels, but it would violate C3).

This PASG is unfinished, n_4 and n_5 are not expanded. n_0 is an example for the remaining constraints. A2) is satisfied, as tracking x in L_a is enough to disable c_3, and both c_1 and c_2 are enabled in L_c and everywhere in L_a. As the outgoing edges are labeled with c_1 and c_2, B1) is satisfied. Let $e = (n_0, c_1, d)$ be the c_1 edge from n_0. The assignments in c_1 are $a_1^{c_1} = (x' := x + 1 \wedge y' := y)$, paired with $n_1^e = n_2$ and $a_2^{c_1} = (x' := x \wedge y' := y)$ paired with $n_2^e = n_1$, which satisfies b2). E.g. for $a_1^{c_1}$: $eval(a_1^{c_1}, L_c(n_0)) = eval((x' := x + 1 \wedge y' :=$

$y), (x = 0, y = 0)) = (x = 1, y = 0), eval(a_1^{c_1}, L_a(n_0)) = (x = 1) \preceq (x = 1)$ and $d_{c_1}(a_1^{c_1}) = 0.8 = d(n_2)$.

Algorithm 1 PASG construction

1: $N \leftarrow \{n_0\}; E_T \leftarrow \{\}; E_C \leftarrow \{\}$; $L_c(n_0) \leftarrow s_0; L_a(n_0) \leftarrow \top$; waitlist $\leftarrow \{n_0\}$
2: **while** waitlist is not empty **do**
3: $\quad$ $n \in$ waitlist ; waitlist $\leftarrow$ waitlist$\setminus\{n\}$
4: $\quad$ **if** $\exists\, n_c \neq n \in N : L_c(n) \in L_a(n_c) \land n_c$ not covered **then**
5: $\quad\quad$ $E_C \leftarrow E_C \cup \{(n, n_c)\}$; $Block(n, \neg L_a(n_c))$
6: $\quad$ **else** $\hfill \triangleright$ *Expansion*
7: $\quad\quad$ **for all** $c \in \mathcal{C}$ **do**
8: $\quad\quad\quad$ **if** $eval(g_c, L_c(n)) = True$ **then** $\hfill \triangleright$ *c is enabled in $L_c(n)$*
9: $\quad\quad\quad\quad$ **if** c is target command **then** mark n as target
10: $\quad\quad\quad\quad$ **if** $eval(g_c, L_a(n)) = Unknown$ **then**
11: $\quad\quad\quad\quad\quad$ $Block(n, \neg g_c)$ $\hfill \triangleright$ *c can be disabled in L_a, so we refine*
12: $\quad\quad\quad\quad$ **for all** $a_i \in Supp(d_c)$ **do**
13: $\quad\quad\quad\quad\quad$ $N \leftarrow N \cup \{n_{new}\}; L_c(n_{new}) \leftarrow eval(a, L_c(n)); L_a(n_{new}) \leftarrow \top;$
$\quad\quad\quad\quad\quad$ $\delta(n_{new}) \leftarrow p_i^c$ waitlist $\leftarrow$ waitlist $\cup\, n_{new}$
14: $\quad\quad\quad\quad$ $E_T \leftarrow E_T \cup (n, c, \delta)$
15: $\quad\quad\quad$ **else if** $eval(g(c), L_a(n)) = Unknown$ **then**
16: $\quad\quad\quad\quad$ $Block(n, g_c)$ $\hfill \triangleright$ *c can be enabled in L_a, but is not in L_c, so we refine*
17: **return** PASG (N, E_T, E_C, L_c, L_a)

Algorithm 2 Block(n, ϕ)

Require: $eval(\phi, L_c(n)) = False$
18: $L_a(n) \leftarrow block(L_a(n), \phi, L_c(n))$
19: **for all** $(n', n) \in E_C$ **do** $\hfill \triangleright$ *Check nodes n' covered by n*
20: $\quad$ **if** $L_c(n') \notin L_a(n)$ **then** $\hfill \triangleright$ *Remove if new L_a cannot cover*
21: $\quad\quad$ $E_C \leftarrow E_C \setminus (n', n); waitlist \leftarrow waitlist \cup\, n'$
22: $\quad$ **else** $Block(n', \neg L_a(n))$ $\hfill \triangleright$ *Else refine covered node*
23: Let $e = (n_{pre}, c, d) \in E_T$ $s.t.$ $n \in Supp(d)$ $\hfill \triangleright$ *n_{pre} denotes the parent of this node*
24: Let a_i^c $s.t.$ $n_i^e = n$ $\hfill \triangleright$ *a_i^c is the assignment which resulted in the node n*
25: $Block(n_{pre}, \neg eval^{-1}(a_i^c, L_a(n)))$ $\hfill \triangleright$ *Making sure that **B2)** still holds for L_a*

Exploration. Algorithm 1 shows PASG construction. An initial node n_0 labeled $L_c(n_0) = val_0, L_a(n_0) = \top$ is extended to a well-labeled PASG with each node either covered or expanded. Algorithm 2 shows blocking an expression from $L_a(n)$, used during refinement.

When removing a node n from the waitlist, we check whether $\exists n_c \neq n \in N : L_c(n') \in L_a(n_c)$ s.t. n_c is not covered. If so, a covering edge (n', n_c) is created and $L_a(n')$ is strengthened for C2) to hold. Else, it is expanded.

If a node $n \in N$ is selected for expansion, we check for each $c \in \mathcal{C}$ whether $eval(g_c, L_c(n)) = True$. If so, a new node n_i' is created for each $a_i \in Supp(d_c)$ with $L_c(n_i') = eval(a_i, L_c(n)), L_a(n_i') = \top$, and a transition edge (n, c, d_e) is created such that $d_e(n_i') = d_c(a_i)$ for $i = 1 \ldots |Supp(d_c)|$. Because of A2), if the abstract label contains states where the transition is disabled, we remove them by blocking out the negated guard (Line 10).

Table 2. PASG constraints

Constraint	Formalisation
A1) Abstract label contains the concrete label:	$\forall n \in N : L_c(n) \in L_a(n)$
A2) Concrete label exactly represents the whole abstract label with respect to the enabled commands	$\forall n \in N : \forall c \in C : eval(g_c, L_c(n)) = eval(g_c, L_a(n))$
B1) The command of a transition edge is enabled in the concrete label of the source	$\forall (n, c, \cdot) \in E_T : eval(g_c, L_c(n)) = True$
B2) For transition edges $e = (n, c, d) \in E_T$, the ith result node is consistent with the ith assignment: same probability, concrete label is the result of the assignment, abstract label overapproximates the result	$d(n_i^e) = d_c(a_i^c)$ $L_c(n_i^e) = eval(a_i^c, L_c(n))$ $eval(a_i^c, L_a(n)) \preceq L_a(n_i^e)$
C1) Abstract label of covering node contains the concrete label of covered node	$\forall (n, n') \in E_C : L_c(n) \in L_a(n')$
C2) Covering node is at least as abstract as the covered node	$\forall (n, n') \in E_C : L_a(n) \preceq L_a(n')$
C3) Covering node is not covered	$\forall (n, n') \in E_C : \neg \exists (n', n'') \in E_C$
D1) At most one node labeled with a given concrete label can be non-covered	$\forall n, n' \in N : n \neq n' \wedge L_c(n) = L_c(n') \implies (\exists n'' : (n, n'') \in E_C \vee (n', n'') \in E_C)$

If $eval(g_c, L_c(n)) = False$, we compute $eval(g_c, L_a(n))$. If $False$, we move on to the next command. If $Unknown$ (cannot be $True$, as $L_c \in L_a$), A2) is violated, so $L_a(n)$ needs to be strengthened: a new abstract label is computed as $\hat{s}' = block(L_a(n), g_c, L_c(n))$. Because of the contract of $block$, $eval(g_c, \hat{s}') = False$, so this command no longer causes a constraint violation (Line 15).

Refinement. Refinement is interleaved with exploring the abstract state space. Whenever the $L_a(n)$ changes for some $n \in N$, the constraints may be violated. If constraint C1) is violated, the problematic covering edge is removed from E_C. This makes n non-covered, so we expand it later (Line 21).

If constraint C2) is violated by covering edge $(n, n') \in E_C$, but constraint C1) still holds, the current $L_a(n)$ must be replaced with $\hat{s}'$ such that $L_c(n) \in \hat{s}'$, $\hat{s}' \preceq L_a(n')$ and $\hat{s}' \preceq L_a(n)$ (referring to the current L_a). An appropriate $\hat{s}'$ is $block(L_a(n), \neg L_a(n), L_c(n))$ (Line 22).

Assume that B2) is violated by an edge $e = (n, c, d)$. Because of how the PASG is constructed, the concrete label and probability subconstraints of B2) must still hold, but the abstract label part is violated by some node n_i^e: L_a of n_i^e no longer overapproximates applying a_i^c (the assignment that led to its creation) to L_a of its parent. The violation caused by this assignment is eliminated by changing $L_a(n)$ to $block(L_a(n), eval^{-1}(a_i^c, L_a(n_i^e)), L_c(n))$ (Line 25).

Strengthenings may create new violations, but all of them are eliminated after finite steps (if the concrete label can be finitely represented in the abstract domain), and we continue expanding non-covered nodes. Efficient implementations of the algorithm can employ sequence interpolation to strengthen the whole path up to the root at once [27], which we do when using the predicate domain, as we observed that both simple weakest-precondition-based refinement and binary interpolation lead to predicates growing very fast.

Example 2. Figure 1b shows an example of refinement in orange. Expanding n_5, we realize c_3 is enabled in some states described by abstract label $x = 2$, but not in the concrete label $x = 2, y = 0$. Thus, we strengthen n_5 by blocking the guard $x == 2 \wedge y == 2$, resulting in the abstract label $x = 2, y = 0$.

This triggers another strengthening, as $L_a(n_5)$ no longer overapproximates applying $x' = x + 1 \wedge y' = y$ to $L_a(n_2)$. n_2 is also strengthened, removing a cover edge as $L_c(n_3)$ is no longer contained in $L_a(n_2)$. Strengthening a covering node also strengthens the covered nodes if the covering remains (see n_4 and n_1).

Numerical Analysis. The finished PASG can be treated as an MDP $(N, \mathcal{C} \cup cover, T_{PASG}, n_0)$. Regarding T_{PASG}, for a non-covered node n, a command $c \in C$ is enabled if there is an edge $(n, c, d) \in E_T$ in the PASG, and $T_{PASG}(n, c) = d$. The only action in covered nodes is *cover*, which results in their covering node.

Theorem 1. *The maximal/minimal probability of reaching a target node in the PASG of an MDP M is the same as in M.*

Refer to the Appendix for proofs[1].

Due to the symmetry of constraint A2), this abstraction can be considered similar to bisimulation-reduction, but not aiming for the coarsest bisimulation. Constructing the PASG can be computationally cheaper than the coarsest bisimulation, but the larger state space may result in more expensive numerical computation. The advantages appear when the abstraction is combined with partial state space exploration, as most bisimulation reduction algorithms in the literature cannot be done on the fly. Comparing our method to bisimulation reduction in depth (both theoretically and empirically) is planned for future work.

3.1   Combining with BRTDP

Bounded Real-Time Dynamic Programming (BRTDP) [3,22] approximates the value function of an MDP *during state space exploration*. It maintains an upper and a lower bound (U and L) by generating traces and updating the bounds for the encountered states. In each step, the optimal action is chosen according to the current U. The strategy for choosing a state from the result distribution is a parameter of the algorithm, for which we implemented the RANDOM and DIFF_BASED trace generation strategies from [3].

Our lazy abstraction algorithm combines well with such methods. As refinement is performed during expansion, the abstract states in an in-progress PASG are coarser than those in a finished PASG. Thus, if BRTDP reaches the required threshold before constructing the full PASG, the abstract labels remain coarser.

---

[1] Available online: http://arxiv.org/abs/2406.00824.

Existing probabilistic CEGAR schemes like [17] cannot benefit from this, as they need lower and upper value bounds for all nodes for refinement, while BRTDP works best when only the value of the initial node is needed. This combination enables a controlled trade-off between time and accuracy.

The skeleton of the algorithm is the same as the BRTDP algorithm described in [3] for general MDPs. The difference is that instead of generating traces from the final PASG, we use the steps of building the PASG for trace generation. As such, refinement is also possible during trace simulation, potentially removing covers used in previous simulations.

Theorem 2. *The maximal probability of reaching a target state is always between $L(n_0)$ and $U(n_0)$ if BRTDP is used with PASG construction steps.*

This theorem is non-trivial, as the traces are not generated for the *finished* PASG, but for an in-progress version where transient cover edges can exist which would not be present when finished. The main idea behind its proof is that if a value is propagated through a cover edge, then the value has been updated only based on traces consistent with the cover edge (as we would have already removed it if any trace inconsistent with it had been explored).

4 Evaluation

Implementation. Our prototype[2] with the explicit value and predicate domains is implemented in the Theta model checker [26], taking JANI models [4] as input. Only properties of the form $P_{max}(p\ U\ q)\ =?$ are supported where p, q are Boolean constraints. Action result probabilities must be constant. The locations of JANI models are always tracked in the abstract states and covering can only occur between nodes with the same locations. To fairly assess the algorithms rather than the implementations, we implemented both Bounded Value Iteration (BVI) [1] and BRTDP as MDP solution techniques in Theta both with and without lazy abstraction. We precompute almost sure reachability and avoidance to speed up the numerical solution if BVI is used.

The main metrics of interest are state space size and running time. As the numerical computations mostly scale with the non-covered PASG nodes, we are especially interested in the number of non-covered nodes. Our evaluation focuses on the following research questions:

RQ1. *How does lazy abstraction affect the state space size and analysis time?*

RQ2. *Does the combination of lazy abstraction with BRTDP lead to reduced abstract model size when converged? How does it affect the running time?*

---

[2] https://github.com/szdan97/probabilistic-theta/tree/prob-proto.

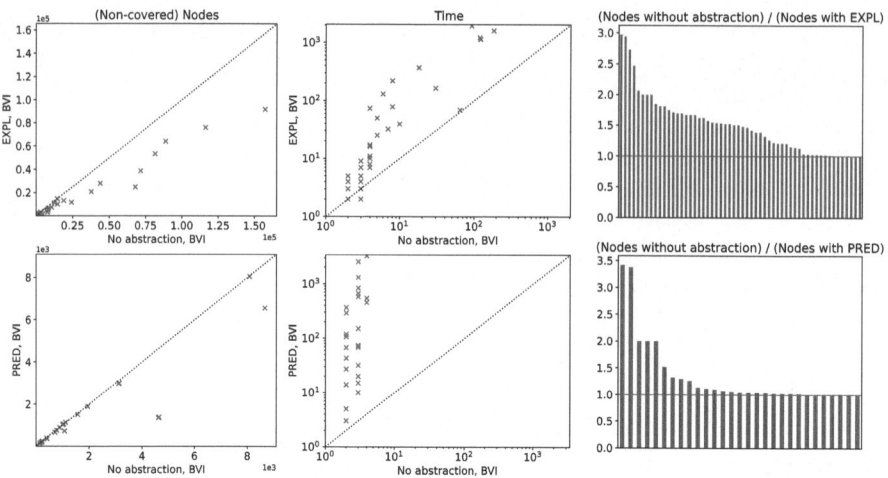

Fig. 2. Comparison of standard and abstract BVI with the EXPL (top row) and PRED (bottom row) domain. Columns: 1st: number of (non-covered) nodes, 2nd: running time on log scale, 3rd: ratio of the original and abstract state space size (red line marks 1, which would be no reduction). Only for inputs where the two algorithms compared in the plot terminated. (Color figure online)

Setup. We used the 104 MDP model-property pairs from the Quantitative Verification Benchmark Set [12] compatible with our current implementation. The experiments were conducted using BenchExec [2], running them on the *Komondor HPC*[3] with AMD EPYC™ 7763 CPUs, each run getting 8 CPU cores, 16GB RAM, and a 1-hour timeout. Convergence threshold was 10^{-6} (absolute) for all algorithms.

Results and Discussion RQ1. Figure 2 shows the BVI results. Lazy abstraction often significantly reduced the state space: for example, a 3-fold reduction was possible for *beb.3-4 [N=3, prop: GaveUp]* (from 4632 nodes to 1559 non-covered nodes) and *csma.2-6 [prop: all_before_max]* (from 67741 to 24837) with EXPL, and for *beb.3-4 [N=3, prop: GaveUp]* (from 4632 to 1354) with PRED. There are also inputs where the explicit domain could not reduce the state space size (e.g. *blocksworld.5, cdrive.2*, which is expected because of its low granularity).

Measuring the analysis time, it turned out that the overhead of more complex operations during exploration outweighed the benefits of numerical computations on a smaller state space. The overhead is apparent for EXPL, but it is much less severe than for PRED.

Predicate abstraction was sometimes able to reduce the state space more than the explicit domain when it terminated before timeout, but the opposite was also present (related plots can be found in the appendix). The overhead of

---

[3] https://hpc.kifu.hu/en.

interpolating using an SMT solver was often too large, and so PRED often failed to terminate in time.

We identified several abstraction-specific optimization possibilities for the implementation. For one, the interpolants returned by Z3 were often very large and had a redundant structure, which we could mitigate through structural simplification, improving both time and memory efficiency. Investigating alternative refinement methods and other solvers could also lead to better results.

Another opportunity for optimization is that like the non-probabilistic version [27], we currently create multiple nodes with the same L_C if a concrete state is reachable on multiple paths. As one of them always covers the others, only that is explored. However, according to our investigation, there are inputs where this led to multiple magnitudes of increase in the number of nodes during exploration. Merging such nodes instead would solve this issue, but refinement must be slightly changed as the path up to the root will no longer be unambiguous. We already had some preliminary measurements with promising results regarding this modification.

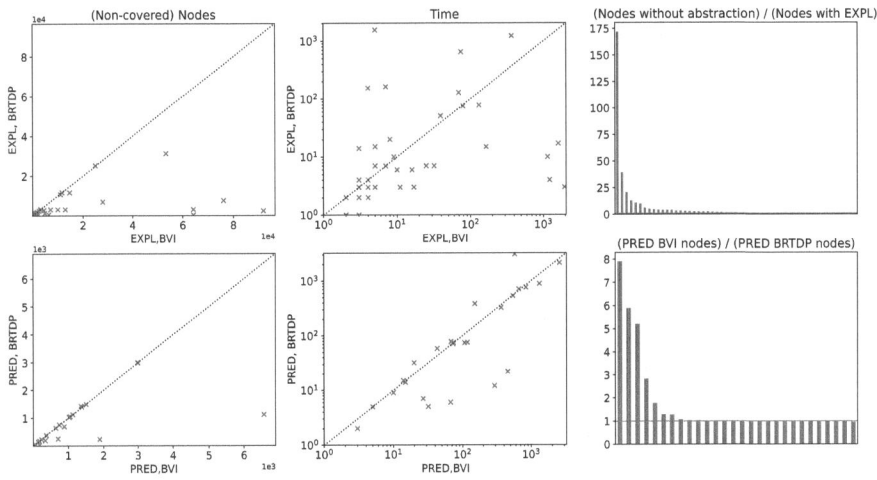

Fig. 3. Comparison of abstract BVI and BRTDP. Same plot types as Fig. 2.

RQ2. The simulation-based nature of BRTDP makes it harder to gauge the benefits of abstract BRTDP compared to standard BRTDP: there were inputs where abstract BRTDP converged with fewer nodes and where the concrete one did. (We relegated plots related to this comparison to the appendix.)

The benefits of abstract BRTDP compared to abstract BVI are much more apparent. The plots in Fig. 3 show this comparison (the results of the strategy leading to fewer nodes were used for each input-domain pair for BRTDP).

The highest relative state space size reduction was on *zeroconf [N=20, k=2, !reset, prop: correct]* (170-fold from 64109 to 373 non-covered nodes) for

EXPL and *pnueli-zuck.3 [prop: live]* (from 1888 to 239 non-covered nodes) for PRED. There were inputs where no further reduction was achieved though (e.g. *blocksworld.5, cdrive.2, rectangle-tireworld.5, ij.10* for both domains). When both BVI and BRTDP terminated, BRTDP was often able to do so in less time, especially with PRED (the results are much more two-sided for EXPL).

5 Conclusions

We proposed a lazy abstraction algorithm for symbolic MDPs and combined it with BRTDP. We provided numerical evaluation for different versions of the proposed algorithm using the Quantitative Verification Benchmark Set, comparing them to explicitly computing the concrete state space.

The initial experimental evaluation shows potential in the proposed algorithm, especially if reducing the state space is paramount for staying within the memory limits. As the time overhead introduced by more complex computations in the state space exploration often outweighed the gains from analyzing a smaller state space, we plan to explore possible improvements for this aspect. Further measurements using other benchmark sets and more parameterizations of the scalable models in the QVBS are also planned.

Additionally, we wish to explore *fine-grained transitioning* between the strict representation version (see A2)) and a more direct overapproximating adaptation by incorporating ideas from *game-based abstraction refinement*, moving the algorithm closer to standard abstraction approaches instead of bisimulation reduction.

References

1. Baier, C., Klein, J., Leuschner, L., Parker, D., Wunderlich, S.: Ensuring the reliability of your model checker: interval iteration for Markov decision processes. In: Majumdar, R., Kunčak, V. (eds.) CAV 2017. LNCS, vol. 10426, pp. 160–180. Springer, Cham (2017). https://doi.org/10.1007/978-3-319-63387-9_8
2. Beyer, D., Löwe, S., Wendler, P.: Reliable benchmarking: requirements and solutions. Int. J. Softw. Tools Technol. Transf. (2019). https://doi.org/10.1007/s10009-017-0469-y
3. Brázdil, T., et al.: Verification of Markov decision processes using learning algorithms. In: Cassez, F., Raskin, J.-F. (eds.) ATVA 2014. LNCS, vol. 8837, pp. 98–114. Springer, Cham (2014). https://doi.org/10.1007/978-3-319-11936-6_8
4. Budde, C.E., Dehnert, C., Hahn, E.M., Hartmanns, A., Junges, S., Turrini, A.: JANI: quantitative model and tool interaction. In: Legay, A., Margaria, T. (eds.) TACAS 2017. LNCS, vol. 10206, pp. 151–168. Springer, Heidelberg (2017). https://doi.org/10.1007/978-3-662-54580-5_9
5. Chadha, R., Viswanathan, M.: A counterexample-guided abstraction-refinement framework for Markov decision processes. ACM TOCL (2010). https://doi.org/10.1145/1838552.1838553
6. Clarke, E., Grumberg, O., Jha, S., Lu, Y., Veith, H.: Counterexample-guided abstraction refinement. In: Emerson, E.A., Sistla, A.P. (eds.) CAV 2000. LNCS, vol. 1855, pp. 154–169. Springer, Heidelberg (2000). https://doi.org/10.1007/10722167_15

7. Cousot, P., Monerau, M.: Probabilistic abstract interpretation. In: Seidl, H. (ed.) ESOP 2012. LNCS, vol. 7211, pp. 169–193. Springer, Heidelberg (2012). https://doi.org/10.1007/978-3-642-28869-2_9

8. D'Argenio, P.R., Jeannet, B., Jensen, H.E., Larsen, K.G.: Reduction and refinement strategies for probabilistic analysis. In: Hermanns, H., Segala, R. (eds.) PAPM-PROBMIV 2002. LNCS, vol. 2399, pp. 57–76. Springer, Heidelberg (2002). https://doi.org/10.1007/3-540-45605-8_5

9. de Alfaro, L., Roy, P.: Magnifying-lens abstraction for Markov decision processes. In: Damm, W., Hermanns, H. (eds.) CAV 2007. LNCS, vol. 4590, pp. 325–338. Springer, Heidelberg (2007). https://doi.org/10.1007/978-3-540-73368-3_38

10. Dehnert, C., Katoen, J.-P., Parker, D.: SMT-based bisimulation minimisation of Markov models. In: Giacobazzi, R., Berdine, J., Mastroeni, I. (eds.) VMCAI 2013. LNCS, vol. 7737, pp. 28–47. Springer, Heidelberg (2013). https://doi.org/10.1007/978-3-642-35873-9_5

11. Esparza, J., Gaiser, A.: Probabilistic abstractions with arbitrary domains. In: Yahav, E. (ed.) SAS 2011. LNCS, vol. 6887, pp. 334–350. Springer, Heidelberg (2011). https://doi.org/10.1007/978-3-642-23702-7_25

12. Hartmanns, A., Klauck, M., Parker, D., Quatmann, T., Ruijters, E.: The quantitative verification benchmark set. In: Vojnar, T., Zhang, L. (eds.) TACAS 2019. LNCS, vol. 11427, pp. 344–350. Springer, Cham (2019). https://doi.org/10.1007/978-3-030-17462-0_20

13. Henzinger, T.A., Jhala, R., Majumdar, R., Sutre, G.: Lazy abstraction. In: POPL 2002 (2002)

14. Herbreteau, F., Srivathsan, B., Walukiewicz, I.: Lazy abstractions for timed automata. In: Sharygina, N., Veith, H. (eds.) CAV 2013. LNCS, vol. 8044, pp. 990–1005. Springer, Heidelberg (2013). https://doi.org/10.1007/978-3-642-39799-8_71

15. Hermanns, H., Wachter, B., Zhang, L.: Probabilistic CEGAR. In: Gupta, A., Malik, S. (eds.) CAV 2008. LNCS, vol. 5123, pp. 162–175. Springer, Heidelberg (2008). https://doi.org/10.1007/978-3-540-70545-1_16

16. Kamaleson, N., Parker, D., Rowe, J.E.: Finite-horizon bisimulation minimisation for probabilistic systems. In: Bošnački, D., Wijs, A. (eds.) SPIN 2016. LNCS, vol. 9641, pp. 147–164. Springer, Cham (2016). https://doi.org/10.1007/978-3-319-32582-8_10

17. Kattenbelt, M., Kwiatkowska, M., Norman, G., Parker, D.: Abstraction refinement for probabilistic software. In: VMCAI 2009 (2009)

18. Kelmendi, E., Krämer, J., Křetínský, J., Weininger, M.: Value iteration for simple stochastic games: stopping criterion and learning algorithm. In: CAV 2018 (2018)

19. Komuravelli, A., Păsăreanu, C.S., Clarke, E.M.: Assume-guarantee abstraction refinement for probabilistic systems. In: Madhusudan, P., Seshia, S.A. (eds.) CAV 2012. LNCS, vol. 7358, pp. 310–326. Springer, Heidelberg (2012). https://doi.org/10.1007/978-3-642-31424-7_25

20. Kwiatkowska, M., Norman, G., Parker, D.: Stochastic model checking. In: SFM 2007 (2007)

21. Kwiatkowska, M., Norman, G., Parker, D.: PRISM 4.0: verification of probabilistic real-time systems. In: CAV 2011 (2011)

22. McMahan, H.B., Likhachev, M., Gordon, G.J.: Bounded real-time dynamic programming: RTDP with monotone upper bounds and performance guarantees. In: ICML 2005 (2005). https://doi.org/10.1145/1102351.1102423

23. McMillan, K.L.: Lazy abstraction with interpolants. In: Ball, T., Jones, R.B. (eds.) CAV 2006. LNCS, vol. 4144, pp. 123–136. Springer, Heidelberg (2006). https://doi.org/10.1007/11817963_14

24. Parker, D., Norman, G., Kwiatkowska, M.: Game-based abstraction for Markov decision processes. In: QEST 2006 (2006). https://doi.org/10.1109/QEST.2006.19

25. Song, L., Zhang, L., Hermanns, H., Godskesen, J.C.: Incremental bisimulation abstraction refinement. ACM TECS (2014). https://doi.org/10.1145/2627352

26. Tóth, T., Hajdu, A., Vörös, A., Micskei, Z., Majzik, I.: Theta: a framework for abstraction refinement-based model checking. In: FMCAD 2017 (2017). https://doi.org/10.23919/FMCAD.2017.8102257

27. Tóth, T., Majzik, I.: Configurable verification of timed automata with discrete variables. Acta Informatica (2022). https://doi.org/10.1007/s00236-020-00393-4

28. Wachter, B., Zhang, L.: Best probabilistic transformers. In: Barthe, G., Hermenegildo, M. (eds.) VMCAI 2010. LNCS, vol. 5944, pp. 362–379. Springer, Heidelberg (2010). https://doi.org/10.1007/978-3-642-11319-2_26

Queueing Analysis of an Ensemble Machine Learning System

Keishin Tsutsumi[1]([✉]), Tuan Phung-Duc[2][iD], and Hong-Linh Truong[3][iD]

[1] Degree Programs in Systems and Information Engineering, University of Tsukuba,
1-1-1 Tennodai, Tsukuba, Ibaraki 305-8573, Japan
`s2420507@u.tsukuba.ac.jp`
[2] Institute of Systems and Information Engineering, University of Tsukuba,
1-1-1 Tennodai, Tsukuba, Ibaraki 305-8573, Japan
`tuan@sk.tsukuba.ac.jp`
[3] Department of Computer Science, Aalto University, Espoo, Finland
`linh.truong@aalto.fi`

Abstract. Recent advances in AI/ML technologies have accelerated the development of various ML applications. One of the major trends in AI/ML application development is the increasing use of multiple ML models to support high-accuracy inference in a complex end-to-end ML serving. However, testing the right configuration of multiple ML models is expensive, and the application requirements for ML inferences are highly dependent on various factors like the quality of ML models, computing resource performance, and data quality. In this context, techniques and methods that help to emulate and analyze ML inference characteristics using queueing theory can reduce the development effort and cost for ML services encapsulating ML models but also the entire ML system. In this paper, we modeled and analyzed a queueing model for an ML system that uses ensemble learning as an inference method with a new rule and clarified the impacts of model design in ensemble learning on the system's performance. As a result, we demonstrate the usefulness of the analysis for understanding possible configurations and their efficiency in the ML system through queueing analysis and simulation.

1 Introduction

In recent years, Machine Learning System (MLS) is becoming familiar with us and applied in various fields. The application of machine learning (ML) has been expanded to automatic driving technology, digital twins, and many other fields [6]. However, even with such technological development, the development of MLS is still costly, and this is a very important issue to the development of MLS in the real world. In recent years, new ML algorithms have been proposed one after another to improve the accuracy of ML inference, while in the field of computing, computing power is growing quickly. Also, testing methods for corner cases have been studied to evaluate the reliability of ML models [1,9]. However, these studies focused on the reliability and safety of the models and

A. Devos et al. (Eds.): ASMTA 2024, LNCS 14826, pp. 97–111, 2025.
https://doi.org/10.1007/978-3-031-70753-7_7

the computers themselves. Therefore, there are still few studies that focus on the structural issues of the MLS as a whole, to help reduce the development effort, which is the subject of this study.

We are interested in studying congestion of ML jobs and delays in system processing to help selecting suitable configurations for ML ensembles used in ML systems. Our approach is to tackle these problems using queueing theory, which assumes that the customer is the ML job and the server is the inference model. There is already a queueing study on MLS using two models, one assumes parallel servers and the other assumes shared servers [4]. Precisely, [4] analyzes the behavior of the model assuming two inputs and two servers theoretically. In [11], the same model as [4] is verified experimentally using actual equipment. Also, [2] analyzes the reliability of the output itself when n versions of the model are introduced. However, the number of models is limited to two in [4,11], and no research has been done in case that the number of models is extended to an arbitrary number n. Besides, in [2], architectural problems such as processing delays have not been discussed.

In this paper, we study MLS with more models than in [2] and provide structural insights into [4,11]. We analyze a queueing model with one input and n models, mimicking an MLS using ensemble learning. Furthermore, we analyze the behavior of the model in case this model is made scalable by the rule that the process is completed when n out of N models have been computed. Throughout this paper, we will identify the impact of the MLS's architecture on the performance of the system. In particular, we investigate the impact of the design parameters n and N on the performance of the system. Our analysis provides a tool for designing a system with targeted performance and reliability.

The rest of our paper is presented as follows. In Sect. 2, we present a concrete real-world application that motivated our model and how our model can help the design of MLS. In Sect. 3, we explain the queueing model. In Sect. 4, we present a detailed analysis of the model and derive the performance measures. In Sect. 5, we show the numerical result for the performance measures and discuss the application to system design. In Sect. 6, we conclude the paper.

2   Motivation

In this section, we present some real-world applications of our model and show how our model can help design these applications.

2.1   A Real-World ML Pipeline

Consider the popular object detection systems. Figure 1 presents a realistic ML pipeline, an MLS, with four main services: *IngestionService*, which receives images/data encapsulated into input requests from clients (like IoT Devices) and stores the images into a storage, serving as a temporary place for holding input data. The status of the input data is stored in a queue from which a *ProcessingService* can claim the request and download the input data to perform some processing tasks. *ProcessingService* will issue inference requests to MLInference-Service which will use ML models to do the classification. Multiple *MLInferanceService* instances will carry the inference, each MLInferenceService may use different ML models. The inference results can be aggregated by using *ResultAggregationService*. In our real development, different interaction patterns among services are used, e.g., *IngestionService* and *ProcessingService* communicates via the request queue with storage as a data buffer, whereas *ProcessingService* calls *MLInferenceService* for inferences.

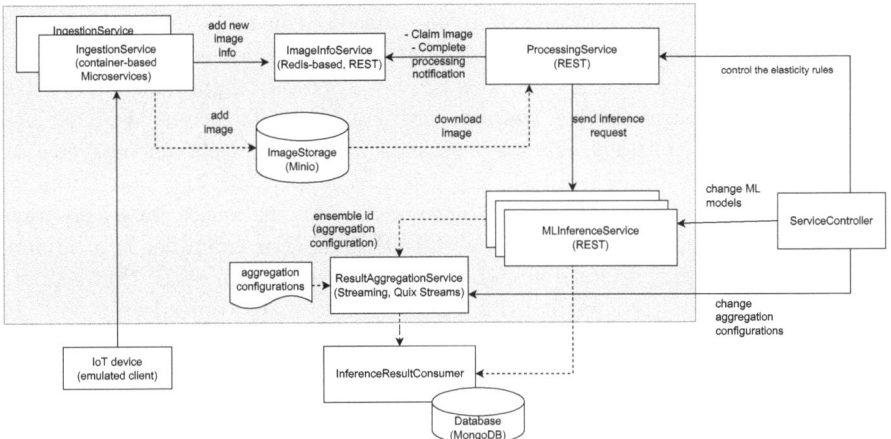

Fig. 1. An example of a real-world pipeline for object classification with multiple ML inference service instances using different ML models. The pipeline, as an MLS, is based on an object classification prototype built and tested with ROHE and detection models from NII [7]. However, the pipeline is also similar to many other ones for object detection and malware detection in terms of ensembles and aggregation [8].

2.2   Supporting MLS Design

In ML engineering, to address the design and optimization of such a pipeline in Fig. 1, a huge effort must be spent on benchmarking ML models and the end-to-end pipelines with different computing resources. This is costly in terms of resources and time consumed and cannot cover many possibilities for the target pipelines very well. On the other hand, using queueing systems and structural analysis could help support some primitive studies and provide guidelines for further development. The main issue is that existing works in queueing-based structured analysis do not reflect the design of realistic pipelines well and lack connection to real-world scenarios.

In this work, we support the developer to carry out prestudy of the deployment configuration (e.g., based on also benchmarking data of individual ML models). For example, the *ServiceController* needs suitable configurations for managing *MLInferenceService* to deliver expected quality for the *InferenceResultConsumer*. We first consider a basic queueing model representing the *ensemble part* of the pipeline – covering *MLInferenceService* and *ResultAggregationService* with realistic ML models. The queueing system consists of c *MLInferenceService* instances in which an ML algorithm with a specific property is utilized. A *job* is an inference request consisting of data to be inferred. Hence *a job* presents the request arriving to *MLInferenceService* that captures data and requests from *IngestionService* to *ProcessingService*. To support the developer, we investigate several policies and compare their performance. For example, we may consider a basic policy that every job needs to be processed in all *MLInferenceService* instances. In another policy, we assume that each job is processed by s among c *MLInferenceInstance*. In this case, jobs can be processed in parallel. Many other possible policies reflect the real characteristics of ML applications. We must also take care of the output aggregation process from the first step. The aggregation process/function is implemented by a service. In reality, this may correspond to comparing outputs from several *MLInferenceService* instances, based on some basic operators like "AND", "OR" or "QUORUM". For example, in a human-face-recognition application, if all ML models have the same result that a photo is human, we adopt that result using AND operation. If we need only one ML model with the output "human", the aggregation process corresponds to OR.

The input data has several important data qualities, such as data completeness and data accuracy, that affect the processing time before the request arrives to *MLInferenceService*. Furthermore, they also affect the ML inference time and inference accuracy [8]. However, in this work, we do not detail them in our queueing analysis. Instead, we attribute such effects to the processing time of the *MLInferenceService* instances. Jobs arrive at the system according to a Poisson process with rate λ. The processing time of all *MLInferenceService* instances, demoted as the servers in the queueing model, follow the same exponential distribution with mean $1/\mu_m$. Each job must be processed in several *MLInferenceService* instances in parallel to achieve higher inference accuracy and confidence to provide reliable result. To build our solution, we also use a real pipeline running in edge-cloud system that leverages tools like QoA4ML[1] and ydata-profiling[2]. We use profiling data from [8] for understanding the real-world systems to support the queueing model design.

3   Models

Given the real-world, common MLS in Fig. 1, we present a queueing model for the MLS. In this queueing model, one job arrives from a *ProcessingService* at the input unit represented by *Buffer*, and the *MLInferenceService* instances perform the inference using an ensemble of multiple trained ML models, the Aggregation Service compares the inference results from each instance (each produced

---

[1] https://github.com/rdsea/QoA4ML.
[2] https://github.com/ydataai/ydata-profiling.

by a ML model), and returns the results to the consumer. ML jobs, due to the arrival of input data, arrive at the system according to a Poisson process with rate λ. An arriving job is processed immediately if no job is waiting in the buffer and the Inference Service instances are waiting. Otherwise, the job enters the Buffer. Jobs are processed by N instances of *MLInferenceService* in parallel, and the inference for an input is completed when n of N instances have been computed. The inference time for each ML model in the inference service follows an exponential distribution with mean $1/\mu_m$. Since the minimum among the $N - k$ random variables that follow the same exponential distribution has mean $1/(N - k)\mu_m$, the average service time for the entire inference service is $\sum_{k=0}^{n-1} 1/(N - k)\mu_m$. If the inference process is finished while the aggregation service is still processing, the inference service enters a stopped state and stops processing until the aggregation process is completed. The time required for processing in the aggregation service follows an exponential distribution with mean $1/\mu_u$. When processing is completed, the aggregation service outputs the results and enters a waiting state, waiting for a job to be sent from the inference service. In practice and real pipelines, we have observed that the processing time for the Aggregation Service is very small, compared with inference times. We present some assumptions of this system model and a description of the random variables for the analysis. First, we assume that the arrival intervals, the inference time and the aggregation time are independent of each other. The ML jobs arriving at the buffer are processed according to the FIFO discipline (first-in first-out). We define, $\mathbb{N}_0 = \mathbb{N} \cup \{0\}$, $\mathbb{S}_m = \{0, 1, 2, \ldots, n, n+1\}$, $\mathbb{S}_u = \{0, 1\}$, $\mathbb{S}_p = \mathbb{S}_m \times \mathbb{S}_u$. Then, $1, \ldots, n$ in $\mathbb{S}_m$ corresponds to the number of *MLInferenceService* instances that must finish computing to complete the inference. Hence, 0 corresponds to "waiting state" which is the state waiting a next job, and $n + 1$ corresponds to "stopped state" which is the state when processing is completed but the Aggregation Service is still in progress. 0 and 1 in $\mathbb{S}_u$ correspond to the two states of "waiting state" and "processing state" for the Aggregation Service. Using these sets, we define the number of ML jobs in the system at time t as $L(t) \in \mathbb{N}_0$ and the state of the *MLInferenceService* and Aggregation Service at the same time as $S(t) \in \mathbb{S}_p \setminus \{(n + 1, 0)\}$. Defining $X(t) = \{L(t), S(t)\}$, then $\{X(t), t \geq 0\}$ is a continuous-time Markov chain on the state space $\mathbb{S}_p$ [3]. It is easy to see that $\{X(t), t \geq 0\}$ forms a quasi-birth-death process, where $L(t)$ and $S(t)$ are the level and the phase, respectively (Fig. 2).

4   Queueing Analysis

In this section, we present the transition diagram and the infinitesimal generator of $\{X(t), t \geq 0\}$ and compute the stationary distribution in Sects. 4.1, 4.2, respectively. In Sect. 4.3, we derive some performance measures of the MLS based on the stationary distribution.

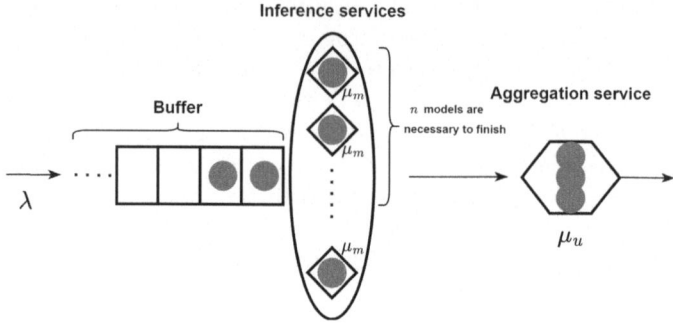

Fig. 2. Queue model for ensemble inference in MLS.

4.1   State Transition Diagram and Infinitesimal Generator

The model can be divided into several cases. First, Fig. 3 is the case that no new ML jobs arrive; it represents only the transitions of states of the inference service and the aggregation service.

In Fig. 3, $i \in \mathbb{S}_m$ is the state of the inference service. 1 to n represents the number of servers remaining to complete the process, 0 represents "waiting state", and $n + 1$ represents "stopped state". Also, $j \in \mathbb{S}_u$ is the state of the Aggregation Service, with 0 and 1 representing "waiting state" and "processing state", respectively. Here, k in Fig. 3 is an integer that satisfies $1 \leq k \leq n$. On the other hand, the transition diagram for each state when a new ML job comes in can be expressed as in Fig. 4.

The infinitesimal generator $\boldsymbol{Q}$ of $\{X(t)\}$ is given as follows. The content of $\boldsymbol{Q}$ is given in the Appendix.

$$
\boldsymbol{Q} = \begin{pmatrix}
q_{0,0} & \boldsymbol{A}_2 & \boldsymbol{O} & \boldsymbol{O} & \cdots & & & \\
\boldsymbol{A}_1 & \boldsymbol{A}_0 & \boldsymbol{C}_2 & \boldsymbol{O} & \boldsymbol{O} & \cdots & & \\
\boldsymbol{O} & \boldsymbol{C}_1 & \boldsymbol{B}_0 & \boldsymbol{B}_2 & \boldsymbol{O} & \boldsymbol{O} & \cdots & \\
\boldsymbol{O} & \boldsymbol{O} & \boldsymbol{B}_1 & \boldsymbol{B}_0 & \boldsymbol{B}_2 & \boldsymbol{O} & \boldsymbol{O} & \cdots \\
\vdots & \boldsymbol{O} & \boldsymbol{O} & \boldsymbol{B}_1 & \boldsymbol{B}_0 & \boldsymbol{B}_2 & \boldsymbol{O} & \boldsymbol{O} \\
& \vdots & \vdots & \ddots & \ddots & \ddots & \ddots & \ddots
\end{pmatrix} .
$$

4.2   Stationary Distribution

We use the information from the previous sections to derive the stationary distribution for the model. First, we define the stationary distribution of this system as follows.

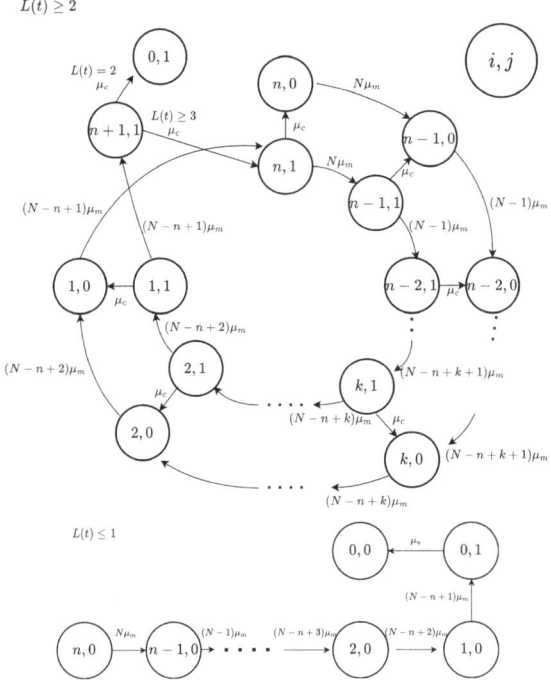

Fig. 3. A diagram of the model (when no new job comes in).

$$\pi_{i,j,k} = \lim_{t \to \infty} P(L(t) = i, S(t) = (j,k)),$$

$$\boldsymbol{\pi}_0 = (\pi_{0,0,0}), \quad \boldsymbol{\pi}_1 = (\pi_{1,0,1} \ \pi_{1,1,0} \ \pi_{1,2,0} \ \cdots \ \pi_{1,n,0}),$$

$$\boldsymbol{\pi}_i = (\pi_{i,1,0} \ \pi_{i,2,0} \ \cdots \ \pi_{i,n,0} \ \pi_{i,1,1} \ \pi_{i,2,1} \ \cdots \ \pi_{i,n+1,1}), \qquad (i \geq 2).$$

$\pi_{i,j,k}$ represents the steady-state probabilities where $L(t) = i$ and $S(t) = (j,k)$, and, $\boldsymbol{\pi}_i$ represents the set of steady-state probabilities with $L(t) = i$. The set of balance equations is given by (1), where $\mathbf{0}$ is a zero row vector with appropriate size according to the context.

$$(\boldsymbol{\pi}_0, \boldsymbol{\pi}_1, \boldsymbol{\pi}_2, \dots) \, \boldsymbol{Q} = \mathbf{0}. \tag{1}$$

Since $\pi_{i,j,k}$ is a probability distribution, the normalization condition is given by (2), where $\boldsymbol{e}_0 = (1), \boldsymbol{e}_1 = \begin{pmatrix} 1 \\ 1 \end{pmatrix}, \boldsymbol{e}_i = \begin{pmatrix} 1 \\ \vdots \\ 1 \end{pmatrix}.$

$$\sum_{k=0}^{\infty} \boldsymbol{\pi}_k \boldsymbol{e}_k = 1. \tag{2}$$

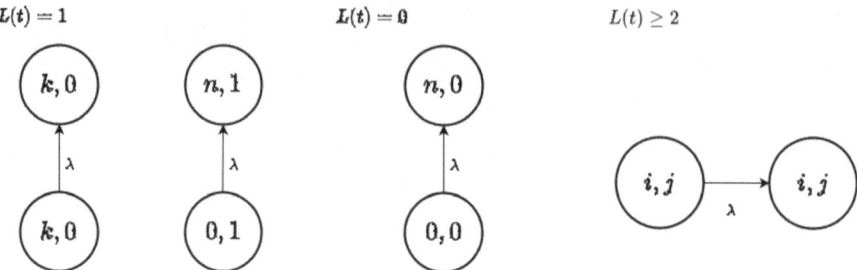

Fig. 4. A diagram of the model (when a new job comes in).

Also, e_i is a matrix of size $(2n + 1) \times 1$ $(i \geq 2)$ with all elements being 1. Here, since Q has the structure of Quasi-Birth-Death process as discussed in the previous sections, we can solve (1), (2) using the method of [10]. In particular, we have

$$\pi_{k+1} = \pi_k R, \; k = 2, 3, 4, \dots , \tag{3}$$

where R is the minimal non-negative solution of

$$B_2 + RB_0 + R^2 B_1 = O. \tag{4}$$

R can be numerically solved using Matrix Analytic Method as in [10]. Therefore, the balance Eq. (1) can be simplified to the following Eqs. (5)–(7).

$$\pi_0 q_{0,0} + \pi_1 A_1 = 0, \tag{5}$$
$$\pi_0 A_2 + \pi_1 A_0 + \pi_2 C_1 = 0, \tag{6}$$
$$\pi_1 C_2 + \pi_2 B_0 + \pi_2 R B_1 = 0. \tag{7}$$

From (2), R, and (3)–(7), we can uniquely determine the stationary distribution π_k $(k = 0, 1, \dots)$.

4.3   Performance Measures

We define performance measures for the model based on the stationary distribution obtained in the previous section. First, let L denote the mean number of jobs in the system. Then we have

$$L = \sum_{k=0}^{\infty} k \pi_k e_k. \tag{8}$$

The average number of jobs in the system represents the expected value of the number of ML jobs in the Buffer, *MLInferenceService*, and Aggregation Service at the moment. Second, the average sojourn time (response time) defined by W is given as follows due to Little's law.

$$W = \frac{L}{\lambda}. \tag{9}$$

The average sojourn time represents the expected time from the arrival of a single job to the completeness of the job at the Aggregation Service. Third, we define the acceptable job arrival rate as follows.

$$
\lambda_B = \boldsymbol{\eta} \begin{pmatrix} 0 \\ \mu_u \\ 0 \\ \mu_u \\ \vdots \\ 0 \\ \mu_c \\ \mu_c \end{pmatrix} / \boldsymbol{\eta} \begin{pmatrix} 1 \\ \vdots \\ 1 \end{pmatrix}.
\tag{10}
$$

where $\boldsymbol{\eta}$ is the stationary distribution of $j \in \mathbb{S}_m$, assuming that the state transitions remain in the homogeneous part of $\boldsymbol{Q}$. Since the system is stable, λ_B is the maximal throughput of jobs of the system. It should be noted that $\lambda < \lambda_B$ guarantees the stability of the system according to [10].

5   Experiments

In this section, we present the numerical results of the performance measures defined in Sect. 4.3. In the numerical results of this paper, we use Monte Carlo simulation and the value of the calculations based on the analytical results explained in the previous sections. The Monte Carlo simulation is to double-check the analytical results derived in Sects. 3 and 4.

In addition, in the cases of Figs. 5, 6, 7 and 8, we assume that the average inference time for each ML model is 0.2 s ($\mu_m = 5$), and the comparison time is 0.02 s ($\mu_u = 50$) to ensure that it is sufficiently short. Also, as for the number of ML models, we assume the case where 1 to 5 specific ML models are enabled out of 10 to 20 models prepared in advance.

5.1   Average Number of Jobs in the System

Figure 5 shows the behavior of the average number of jobs L against n, while $N = 15$, $\lambda = 10, 11, 12$, $\mu_m = 5$, and $\mu_u = 50$. Figure 6 shows the behavior of the average number of jobs L against N, while $n = 5$, $\lambda = 10, 11, 12$, $\mu_m = 5$, and $\mu_u = 50$.

- Theoretical and simulated values of L are almost the same.
- The value of L exponentially increases as the value of n increases.
- The value of L exponentially decreases as the value of N increases.

From the observations in Figs. 5 and 6, we gain some insights on n, N for L. The first is that the number of ML models required to complete the process, i.e., n, has a negative effect on the average number of jobs in the system. The second

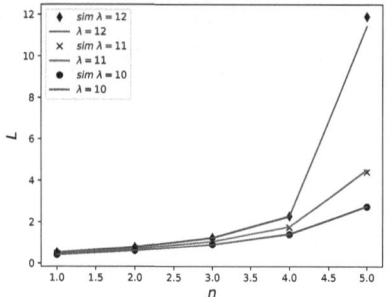

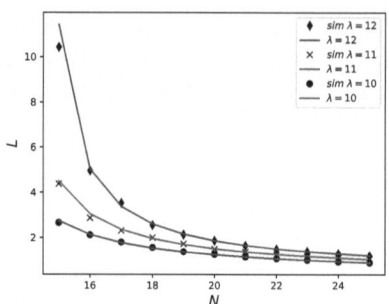

Fig. 5. Average number of jobs in the system by varying the value of n.

Fig. 6. Average number of jobs in the system by varying the value of N.

is that the number of ML models in the system, i.e., N, has a positive effect on the average number of jobs in the system.

From this result, if we want to have a small number of jobs in the system, we may either decrease the number of required ML models (n) or increases the number of ML models in the system (N). However, decreasing n may cause the deterioration of the reliability while increasing N needs additional cost. Thus, using this result, one can find optimal values of n and N given constraints on reliability and cost.

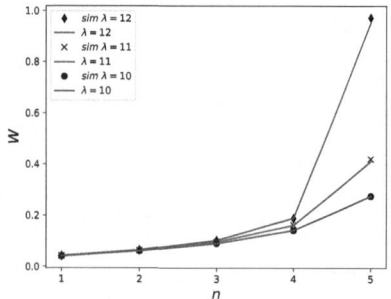

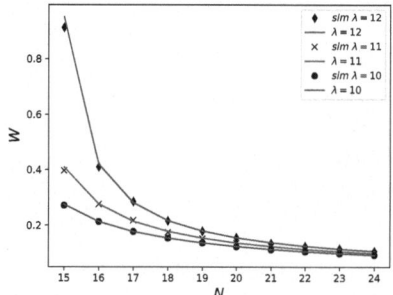

Fig. 7. Average sojourn time by varying the value of n.

Fig. 8. Average sojourn time by varying the value of N.

5.2   Average Sojourn Time

Figure 7 shows the behavior of the average sojourn time against n, while $N = 15$, $\lambda = 10, 11, 12$, $\mu_m = 5$, and $\mu_u = 50$. Figure 8 shows the behavior of the average sojourn time W against N, while $n = 5$, $\lambda = 10, 11, 12$, $\mu_m = 5$, and $\mu_u = 50$.

From the observations in Figs. 7 and 8, we gain the same insights on n, N for W as for L. This result indicates that, if we consider the actual configuration of the system, when the response time performance is deteriorating because of increasing arrival rate, it can be suppressed by adjusting n and N. For example, consider the case of $n = 5$ in Fig. 7: when λ increases from 10 to 12, the processing time increases by a factor of about four.

In this case, reducing n to 4 would reduce the average sojourn time by a factor of about five. In addition to these observations, through Figs. 5, 6, 7 and 8, simulation values and theoretically calculated values are in general agreement. This validates our results.

5.3   Acceptable Job Arrival Rate

Figure 9 shows the behavior of λ_B against n, while $N = 15$, $\mu_m = 1, 3, 5, 7, 10$, and $\mu_u = 50$. Figure 10 shows the behavior of λ_B against N, $n = 15$, $\mu_m = 1, 3, 5, 7, 10$, and $\mu_u = 50$.

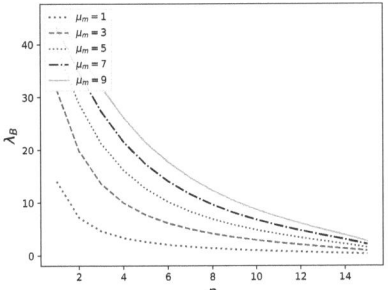

Fig. 9. Acceptable job arrival rate by varying the value of n.

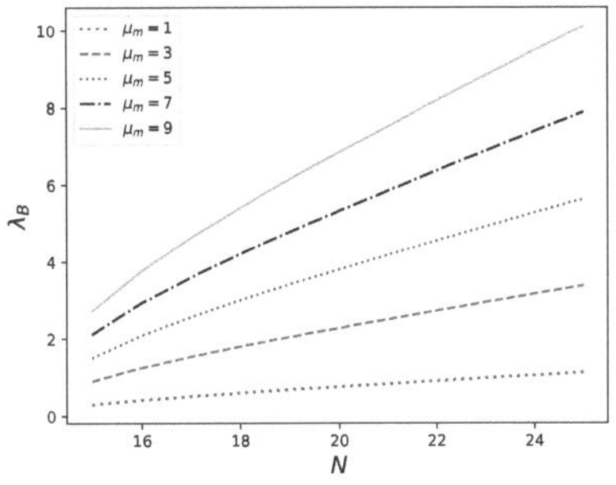

Fig. 10. Acceptable job arrival rate by varying the value of N.

In Figs. 9 and 10, the value of λ_B decreases as the value of n increases, and increases as the value of N increases. Besides, the higher the value of μ_m, the more sensitive the value of λ_B is, to decreasing n and increasing N.

As defined earlier in Sect. 4.3, λ_B represents the maximum throughput, so this result is useful to check the capacity of the ensemble system. In other words, n and N should be adjusted to improve the throughput when λ approaches λ_B.

5.4   Discussion on Supporting the Development of MLS

A practical application of the research results in this paper would be a MLS that can flexibly vary N and n in the ensemble part consisting of multiple *MLInferenceService* of Fig. 1. This variability can improve processing efficiency by varying them during job congestion. For example, in the cases of $\lambda = 12$ in Figs. 5 and 7, by changing n to 4 from 5, L and W decrease by about 1/5, and then λ_B also rises in Fig. 9. On the other hand, when varying N and n to improve processing speed, the trade-off relationship with the output's accuracy and power consumption should be kept in mind. In the ensemble composition, the number and diversity of ML models benefit the accuracy of the output [5]. In other words, when considering reducing n to improve processing efficiency, as in this paper, you must also consider the output error that would increase as a sacrifice. Similarly, when N is increased, the increased power consumption and other costs must be considered. In future work, we would also like to consider the improvement of reliability by increasing n as well as the effect of N on power consumption.

6 Conclusion

In this paper, we proposed a queueing model for MLS that considers ensemble learning as a method of inference and can be extended by adding the original rule, and analyzed it by modeling with a queue. In the model analysis, we used a quasi-birth-death process to compute the stationary distribution of the underlying Markov chain of the model. We also defined performance measures for the model based on the stationary distribution and observed the behavior of the performance measures by both numerical and simulation results.

Our findings from the numerical results are as follows. First, by increasing N and decreasing n, we can improve the system's processing efficiency. This improvement behavior can produce significant improvements up to a certain point, but after that, the improvement will level off. Second, the higher the computational speed of each model, i.e., the higher the value of μ_m, the more sensitive it is to the aforementioned improvement actions.

In future work, we would like to fine-tune queueing parameters based on our real MLS benchmark data. We will compare queueing analysis results with some real tests in our real deployment. We also plan to conduct research that considers the trade-off relationship with processing efficiency, such as prediction accuracy and power consumption.

Acknowlegements. The authors would like to thank Thao-Nguyen Vuong and Minh-Tri Nguyen for the work on the realistic pipeline ML system (the architecture, implementation and observability) used in the motivation illustrated through the paper. This work was supported by JSPS KAKENHI No. 21K11765 and by F-MIRAI: R & D Center for Frontiers of MIRAI in Policy and Technology, the University of Tsukuba and Toyota Motor Corporation Collaborative R & D center.

A   Appendix

A_0 is a matrix of size $(n + 1) \times (n + 1)$ that represents the transition of the number of ML jobs in the system from 1 to 1. Each element $(A_0)_{i,j}$ is defined as follows.

$$(A_0)_{i,j} = \begin{cases} -\mu_u - \lambda & i = 0, \ j = 0, \\ -(N - n + k)\mu_m - \lambda & i = j = k, \ k \in \{1, 2, \ldots, n\}, \\ (N - n + k)\mu_m & i = k, \ j = k - 1, \ k \in \{1, 2, \ldots, n\}, \\ 0 & (otherwise). \end{cases}$$

A_1 is a matrix of size $(n + 1) \times 1$ that represents the transition of the number of ML jobs in the system from 1 to 0. Besides, A_2 is a matrix of size $1 \times (n + 1)$ that represents the transition of the number of ML jobs in the system from 0 to 1. Each element A_1, A_2 is defined as follows.

$$A_1 = \begin{pmatrix} \mu_u \\ 0 \\ \vdots \\ 0 \end{pmatrix}, \qquad A_2 = \begin{pmatrix} 0 \ldots 0 \ \lambda \end{pmatrix}.$$

B_0 is a matrix of size $(2n + 1) \times (2n + 1)$ that represents the transition of the number of ML jobs in the system from n to n, $n \geq 2$. Each element B_0 is defined as follows.

$$(B_0)_{i,j} = \begin{cases} -m_k - \lambda & i = j = 2k, \\ -m_k - \lambda - \mu_c & i = j = 2k + 1, \\ m_k & (i, j) = (2k, \ 2k - 2), \ (2k + 1, \ 2k - 1) \\ -\lambda - \mu_u & i = j = 2n, \\ 0 & (otherwise), \end{cases}$$

where $k = 0, 1, \ldots, n - 1$, $m_k = -(N - n + k + 1)\mu_m$.
B_1 is a matrix of size $(2n + 1) \times (2n + 1)$ that represents the transition of the number of ML jobs in the system from n to $n - 1$, $n \geq 3$. Each element B_1 is defined as follows.

$$(B_1)_{i,j} = \begin{cases} \mu_u & i = 2k + 1, \ j = 2k, \\ \mu_u & i = 2n, \ j = 2n - 1, \\ 0 & (otherwise), \end{cases}$$

where $k = 0, 1, \ldots, n - 1$.
B_2 is a matrix of size $(2n + 1) \times (2n + 1)$ that represents the transition of the number of ML jobs in the system from n to $n + 1$, $n \geq 2$. Each element B_2 is defined as follows.

$$(B_2)_{i,j} = \begin{cases} \lambda & i = j = k, \\ 0 & (otherwise), \end{cases}$$

where $k = 0, 1, \ldots, 2n$.
C_1 is a matrix of size $(2n + 1) \times (n + 1)$ that represents the transition of the number of ML jobs in the system from 2 to 1. Each element C_1 is defined as

follows.

$$C_1 = \begin{pmatrix} 0 & 0 & 0 & \dots & 0 \\ 0 & \mu_u & 0 & \dots & 0 \\ 0 & 0 & 0 & \ddots & 0 \\ 0 & 0 & \mu_u & \ddots & 0 \\ \vdots & \vdots & \ddots & \ddots & 0 \\ \vdots & \vdots & \ddots & \ddots & 0 \\ 0 & \dots & \dots & 0 & 0 \\ 0 & \dots & \dots & 0 & \mu_u \\ \mu_u & 0 & \dots & 0 & 0 \end{pmatrix}.$$

C_2 is a matrix of size $(n + 1) \times (2n + 1)$ that represents the transition of the number of ML jobs in the system from 1 to 2. Each element C_2 is defined as follows.

$$C_2 = \begin{pmatrix} 0 & 0 & \dots & \dots & \dots & 0 & \lambda & 0 \\ \lambda & 0 & 0 & 0 & \dots & \dots & 0 & 0 & 0 \\ 0 & 0 & \lambda & 0 & \ddots & \ddots & \dots & 0 & 0 & 0 \\ \vdots & \vdots & \ddots & \ddots & \ddots & \ddots & \ddots & 0 & 0 & 0 \\ \vdots & \vdots & \vdots & \vdots & \ddots & \ddots & \lambda & 0 & 0 & 0 \\ 0 & 0 & \dots & \dots & \dots & 0 & 0 & \lambda & 0 \end{pmatrix}.$$

References

1. Goodfellow, I., Shlens, J., Szegedy, C.: Explaining and harnessing adversarial examples. In: International Conference on Learning Representations (2015). http://arxiv.org/abs/1412.6572
2. Machida, F.: N-version machine learning models for safety critical systems. In: 2019 49th Annual IEEE/IFIP International Conference on Dependable Systems and Networks Workshops (DSN-W), pp. 48–51 (2019). https://doi.org/10.1109/DSN-W.2019.00017
3. Makimoto, N.: Machigyouretu arugorizumu - Gyouretu kaiseki apurochi -. Asakura Publishing Co., Ltd, Tokyo (2001)
4. Makino, Y., Phung-Duc, T., Machida, F.: A queueing analysis of multi-model multi-input machine learning systems. In: 2021 51st Annual IEEE/IFIP International Conference on Dependable Systems and Networks Workshops (DSN-W), pp. 141–149 (2021). https://doi.org/10.1109/DSN-W52860.2021.00033
5. Mienye, I.D., Sun, Y.: A survey of ensemble learning: concepts, algorithms, applications, and prospects. IEEE Access 10, 99129–99149 (2022). https://doi.org/10.1109/ACCESS.2022.3207287
6. Mori, T., Hayashi, H.: Nihon no chatgpt riyoudoukou (2023 nen 6 gatsu ikou) jakunensou wo chuushinn ni riyouritu ga takamaru. https://www.nri.com/jp/knowledge/report/lst/2023/cc/0622_1, accessed 7 April 2024
7. Nguyen, M.T., Truong, H.L., Arcaini, P., Ishikawa, F.: Optimizing multiple consumer-specific objectives in end-to-end ensemble machine learning serving (2024), under submission

8. Nguyen, M.T., Truong, H.L., Truong-Huu, T.: Novel contract-based runtime explainability framework for end-to-end ensemble machine learning serving. In: International Conference on AI Engineering: Software Engineering for AI, CAIN 2024 (2024). https://conf.researchr.org/dates/cain-2024

9. Ouyang, T., Isobe, Y., Marco, V.S., Ogata, J., Seo, Y., Oiwa, Y.: Ai robustness analysis with consideration of corner cases. In: 2021 IEEE International Conference on Artificial Intelligence Testing (AITest), pp. 29–36 (2021). https://doi.org/10.1109/AITEST52744.2021.00016

10. Takine, T.: M/m/1 wo koete - junshusshoushimetukatei heno shoutai -. In: Operations Research as a Management Science Research, vol. 59, pp. 179–184 (2009)

11. Wakigami, K., Machida, F., Phung-Duc, T.: Reliability and performance evaluation of two-input machine learning systems. In: 2023 IEEE 28th Pacific Rim International Symposium on Dependable Computing (PRDC), pp. 278–286. IEEE Computer Society, Los Alamitos, CA, USA, October 2023.https://doi.org/10.1109/PRDC59308.2023.00044

Analysis of Load Balancing Prioritization for Heterogeneous M/M/c/K Server Clusters in the Stationary Mean-Field Regime

Illés Horváth[1,2(✉)] and Márton Mészáros[3]

[1] HUN-REN-BME Information Systems Research Group, Budapest University of Technology and Economics, Budapest, Hungary
[2] Department of Networked Systems and Services, Budapest University of Technology and Economics, Budapest, Hungary
horvath.illes.antal@gmail.com
[3] Department of Stochastics, Mathematics Institute, Budapest University of Technology and Economics, Budapest, Hungary

Abstract. We examine heterogeneous M/M/c/K server clusters with various load balancing policies. We provide a mathematical framework that allows the efficient computation of the mean system time of jobs in the stationary mean-field limit that allows to find optimal load balancing prioritization.

1 Introduction

For large scale service systems, where service resources (e.g. computing capacity) are distributed to several service units, load balancing plays a crucial role in distributing the total load of the system to ensure better overall service for the incoming tasks (jobs).

There are many different types of load balancing principles. Static load balancing does not take into account the state of the system, instead aiming for a balanced distribution based purely on the incoming jobs. Static load balancing is in general easy to set up, requires minimal overhead communication and performs well when the incoming jobs have some regular patterns.

However, in most systems the incoming jobs have some level of random variability. This situation is generally better handled by load balancing policies which take into account the current state of the system. Scheduling decisions may be based on different types of information, depending on what is available. In general, one of the most important parameters is the current load of the servers, as it is generally desirable to maintain a balanced load among all servers. If available, further information taken into account may include any of the following:

- the server cluster may be heterogeneous, with faster and slower servers;
- job sizes may be used to compute current server load more precisely;

A. Devos et al. (Eds.): ASMTA 2024, LNCS 14826, pp. 112–131, 2025.
https://doi.org/10.1007/978-3-031-70753-7_8

- job and server types may be important in case the servers are heterogeneous and certain servers can serve certain types of jobs more efficiently;
- in some cases, physical location may play a role;
- there may be bottlenecks other than computing capacity in the system (e.g. bandwidth).

In many real-life systems, such information may not be available, but even if it is, there is a tradeoff: a complicated load balancing policy that requires too much communication and computation may generate a significant overhead cost, slowing down the entire system. Hence it is in general desirable to stick to simple load balancing policies.

One of the classic load balancing policies is Join-Shortest-Queue (JSQ), where the incoming job is assigned to the server with the shortest queue (lowest number of jobs) [10]. JSQ offers very even balancing for homogeneous server clusters. However, when the server cluster is heterogeneous, it may be possible to improve upon it. Section 5 shows several such examples.

In order to be able to analyze server clusters, we provide a high-level mathematical framework that allows for calculations in the stationary mean-field limit. The mean-field limit is obtained when the system load and the number of servers scale proportionally to infinity; in the stationary mean-field limit, time also goes to infinity. The mean-field limit has been examined and identified for many different types of systems (not only queuing theory). Typically, the mean-field limit is deterministic and the transient evolution of the system is governed by a system of differential equations. The stationary mean-field limit corresponds to the single stable attractor of this system.

The main performance measure of focus is the mean service time of jobs in the stationary mean-field limit. In order to compute the mean-field stationary distribution, standard computation techniques need to be adapted; to the best of our knowledge, these adapted versions are novel. From the mean-field stationary distribution, computing the mean service time is straightforward via Little's law.

Rigorous proofs are not the main focus of the paper. We do refer to related rigorous results from the literature in cases where they are available, but only provide heuristic arguments for the model examined in the present paper. That said, numerical analysis does support the heuristic computations of the paper.

The paper has a natural precursor in [7], where the general framework is already presented (and in more detail). The derivation of the mean-field stationary distribution and mean system time are novel to this paper, as is the optimization procedure.

The rest of the paper is structured as follows: the rest of this section is dedicated to an overview of load balancing in the literature, and to the necessary mathematical background in population processes (necessary for the mean-field limit). Section 2 describes the general mathematical setup of the server cluster we are interested in. Section 3 contains the main contributions: general description of the behaviour in the stationary mean-field limit, derivation of the mean-field stationary distribution and computation of the mean system time of jobs. Section 4 contains an optimization procedure based on the previous computations, Sect. 5 provides some numerical results, and Sect. 6 concludes the work.

1.1   Load Balancing Principles

Apart from the classic JSQ, several variants have been in use: for JSQ(d), the incoming job is scheduled to the shortest queue from among d servers, selected at random. This offers less balanced load distribution, but also requires less communication. $d = 1$ corresponds to random assignment with no load balancing, and d equal to the total number of servers corresponds to JSQ; as d is increased, it offers better balancing but also more overhead communication. Interestingly, already for $d = 2$, the resulting load balancing policy has certain asymptotic optimality properties [11], often referred to as the power-of-2 (or power-of-d) policies. As a consequence, d is often selected relatively low, such as $d = 2$ or $d = 5$.

The communication overhead cost of JSQ (and similar versions such as JPSQ introduced in Sect. 3 is relatively high since the dispatcher needs to be aware of all queue length during dispatch, while JSQ(d) needs to be aware of only the d selected queue lengths.

For Join-Idle-Queue (JIQ), the incoming job is scheduled to an idle server at random; if there are no idle servers, the assignment is random among all servers. Once again, this offers less balanced load distribution and less communication overhead than JSQ, but, similar to JSQ(d), has some nice asymptotic optimality properties. Mean-field analysis has been carried out for JIQ in [12].

Another related load balancing policy is Join-Below-Threshold (JBT), which associates a threshold with each server; servers below their threshold are considered available and servers above or at their threshold are full. Tasks will be dispatched to a server randomly from among all available servers. This policy again offers less balancing than JSQ, but still offers protection against overloaded servers, and requires communication only when a server switches between available and full. For a full mean-field analysis and cluster optimization of JBT, we refer to [2].

1.2   Density-Dependent Population Processes

In this section, we present mathematical background and framework for density-dependent Markov population processes.

A density-dependent Markov population process has N interacting components, each of which is in a state from a finite set of local states S. The global state of the system is defined as the total number of individuals in each state, that is, a vector $X^N \in \{0, 1, \ldots, N\}^{|S|}$ with $X_1^N + \cdots + X_{|S|}^N = N$. The normalized global state of the system can be defined

$$x^N = \frac{X^N}{N},$$

so $x^N \in [0, 1]^S$ with $x_1^N + \cdots + x_{|S|}^N = 1$.

Each component acts as a continuous time Markov chain. The rate of the transition from $i \in S$ to $j \in S$ is r_{ij}^N (for $i \neq j$). The rates are assumed to be *density-dependent*, that is

$$r_{ij}^N = r_{ij}(x)$$

for some function $r_{ij} : [0,1]^{|S|} \to [0, \infty]$. x is the $|S|$ dimensional vector with elements x_i^N for $i \in \{1, 2, \ldots, |S|\}$. In the classic setup defined by Kurtz [8,9], the functions r_{ij} are usually assumed to be Lipschitz-continuous and independent of N. With this setup, $x^N(t)$ is a continuous time Markov-chain. We define the *mean-field equation* of the system as the following:

$$\frac{\mathrm{d}}{\mathrm{d}t} v_i(t) = \sum_{j \in S} v_j(t) r_{ji}(v(t)), \quad i \in S, \tag{1}$$

where

$$r_{ii} := - \sum_{j \in S, j \neq i} r_{ij}.$$

Lipschitz-continuity guarantees existence and uniqueness of the solution of (1). The following result of Kurtz states mean-field convergence in the transient regime [5,8,9]:

Theorem 1 (Transient mean-field convergence). *Assuming r_{ij} $(i, j \in S)$, are Lipschitz-continuous and*

$$x_i(0) \to v_i(0) \quad i \in \{0, 1, \ldots, |S|\}, \quad \text{in probability,}$$

then for any $T > 0$ we have

$$\lim_{N \to \infty} P \left(\sup_{t \in [0,T]} \|\bar{\mathbf{x}}^N(t) - \mathbf{v}(t)\| > \epsilon \right) = 0.$$

We also have stationary mean-field convergence.

Theorem 2 (Stationary mean-field convergence). *Given the following assumptions:*

- *r_{ij} are Lipschitz-continuous,*
- *the Markov process $x^N(t)$ has a unique stationary distribution π^N for each N, and*
- *(1) has a unique stable attractor $\nu = (\nu_1, \ldots, \nu_{|S|})$,*

we have that the probability measure π^N on S converges in probability to the Dirac measure concentrated on ν.

From Theorem 2 it also follows that

$$E(\pi^N) \to \nu,$$

so ν can be used as an approximation for $E(\pi^N)$ for large N. $E(\pi^N)$ here is basically an $|S|$-dimensional vector of distributions, which converges to a constant

$|S|$-dimensional vector, which can be interpreted as a distribution on S, and is the stable attractor ν.

Theorems 1 and 2 have been generalized further in several ways during recent years. Benaïm and Le Boudec elaborated a framework applicable for a wider range of stochastic processes, which also allows the r_{ij} functions to have a mild dependency on N [3].

The condition on Lipschitz-continuity can also be weakened. For discontinuous r_{ij}'s, (1) turns into a differential inclusion. A formal setup for differential inclusions is quite technical in general, which is omitted from the present paper. For a fully detailed setup, we refer to [6], specifically Theorems 4 and 5, and [13], Theorem 3.5 and Corollary 3.9 for a corresponding version of Theorem 1. [4,14] include some related applications with a discontinuous setting where the corresponding version of Theorem 1 is proved.

For a corresponding version of Theorem 2, we refer to [6], where the main additional condition is that the unique attractor ν lies inside a domain where the r_{ij} are continuous.

For the model of the present paper, the r_{ij} functions will be discontinuous at ν, and identifying the mean-field limit requires understanding the behaviour of the system at discontinuities. Nevertheless, we identify the mean-field limit (based on non-rigorous arguments), and we will also demonstrate mean-field convergence via simulations.

2   Server Clusters

Server clusters examined in this paper are systems of N queues of type M/M/c/K with a common dispatcher.

The job arrival rate to the dispatcher is independent of its state and is constant, with its value being $N\lambda$ (that is, the average arrival rate is λ per server). The dispatcher forwards each incoming arrival to one of the servers according to a load balancing policy (to be described shortly).

In this paper, the term server will generally refer to a single queue; each server has a fixed number of *threads* within each server that act as the service units.

The cluster may have L different server types. Servers within each type $\ell \in \{1, \ldots, L\}$ are identical. The number of threads of a type ℓ server is denoted by $c^{(\ell)}$, and the maximal buffer size is denoted by $K^{(\ell)}$. We assume a finite buffer; for finite buffer sizes, the number of possible prioritization schemes is also finite (see Sect. 3 for prioritization schemes), which helps with optimization (Sect. 4). Also, depending on the load baalncing principle used and the system load (see also the stability condition (3)), buffers will typically be used only up to a certain level, beyond which further buffer size has no effect. We might also occasionally refer to the queues as M/M/c, omitting the buffer size.

Each thread within a server can serve a single job at a time with service rate $\mu^{(\ell)}$. If a type ℓ server holds more than $c^{(\ell)}$ jobs, then $c^{(\ell)}$ jobs are in service at the $c^{(\ell)}$ threads, and the remaining jobs wait in the buffer of that server in a

FIFO manner, and enter a thread immediately as they become available. Jobs currently being served by the threads also count towards the buffer size.

Jobs dispatched to a queue with a full buffer are lost without entering the queue (this will be referred to as data loss).

The total service rate of a server thus depends on the number of jobs i it currently holds, and is simply

$$\mu_i^{(\ell)} = \begin{cases} i\mu^{(\ell)} & \text{for } 1 \leq i \leq c^{(\ell)}, \\ c^{(\ell)}\mu^{(\ell)} & \text{for } c^{(\ell)} < i \leq K^{(\ell)}. \end{cases} \tag{2}$$

We assume L is fixed (independent from N), and the number of servers of type ℓ is denoted by $N^{(\ell)}$, so that

$$N = N^{(1)} + \cdots + N^{(L)}.$$

We are interested in server clusters of different N sizes and especially the limit object as $N \to \infty$, referred to as the mean-field limit (in accordance with Sect. 1.2).

The server cluster is a density-dependent population process, where the current state of a server is simply the number of jobs in its queue. The global state will be denoted by

$$X_i^{(\ell),N}(t), \qquad \left(0 \leq i \leq K^{(\ell)}, \qquad 1 \leq \ell \leq L\right),$$

where $X_i^{(\ell),N}(t)$ is the number of servers with i jobs in its queue at time t. We will mostly use its normalized version

$$x^N(t) = x_i^{(\ell),N}(t), \qquad \left(0 \leq i \leq K^{(\ell)}, \qquad 1 \leq \ell \leq L\right),$$

where

$$x_i^{(\ell),N}(t) = \frac{X_i^{(\ell),N}(t)}{N}.$$

The ratio of each server type is denoted by

$$\gamma_\ell^N = \frac{N_\ell}{N}, \qquad \ell = 1, \ldots, L.$$

γ_ℓ^N may depend on N, but we will assume they converge to some fixed values γ_ℓ as $N \to \infty$. Note that this also means that each $N^{(\ell)}$ goes to infinity, that is, none of the queue types become negligible.

We want the cluster to be stable, so we assume

$$N\lambda < \sum_{\ell=1}^{L} N^{(\ell)} c^{(\ell)} \mu^{(\ell)} \qquad \Longleftrightarrow \qquad \lambda < \sum_{\ell=1}^{L} \gamma_\ell c^{(\ell)} \mu^{(\ell)}. \tag{3}$$

Actually, due to the finite buffer size assumption, the cluster is technically always stable, but we will nevertheless assume (3). In case (3) holds, assuming finite buffer sizes is mostly for technical simplicity, as buffer sizes can be selected large enough to not play a practical role. We also note that (3) is only valid if all L queue types are actually used, which may depend on the load balancing principle; otherwise, the sum on the right should include only the queue types actually used.

The evolution of $x^N(t)$ can be defined formally using Poisson representation. Let

$$P_{i\to(i+1),\ell}(t), \quad 0 \le i \le K^{(\ell)} - 1$$
$$P_{i\to(i-1),\ell}(t), \quad 1 \le i \le K^{(\ell)}$$

denote independent Poisson processes with rate 1. $P_{i\to(i+1),\ell}(t)$ corresponds to arrivals to queues of type ℓ with length i, and $P_{i\to(i-1),\ell}(t)$ corresponds to jobs leaving queues of type ℓ with length i.

The Poisson representation of $x^N(t)$ is

$$
\begin{aligned}
x_i^{(\ell),N}(t) = {}&\frac{1}{N}P_{(i-1)\to i,\ell}\left(N\int_0^t \lambda f_{i-1}^{(\ell)}(x^N(s))\mathrm{d}s\right) \\
&- \frac{1}{N}P_{i\to(i+1),\ell}\left(N\int_0^t \lambda f_i^{(\ell)}(x^N(s))\mathrm{d}s\right) \\
&+ \frac{1}{N}P_{(i+1)\to i,\ell}\left(N\int_0^t \mu_{i+1}^{(\ell)}x_{i+1}^{(\ell),N}(s)\mathrm{d}s\right) \\
&- \frac{1}{N}P_{i\to(i-1),\ell}\left(N\int_0^t \mu_i^{(\ell)}x_i^{(\ell),N}(s)\mathrm{d}s\right),
\end{aligned}
\tag{4}
$$

where $f_i^{(\ell)}(x^N(t))$ denotes the probability that a new arriving job will be dispatched to a queue with length i of type ℓ.

The

$$\{f_i^{(\ell)}(x^N(t))) : 0 \le i \le K_\ell, \ \ell = 1, \ldots, L\}$$

are going to be collectively called the *dispatch functions*. The $f_i^{(\ell)}$ functions correspond directly to the load-balancing principle, which will be addressed later in this section.

Formally, $f_i^{(\ell)}$ are functions defined on the normalized state $x^N(t)$, which are all contained in the compact domain

$$\{x : x \in \mathbb{R}^{\sum_{\ell=1}^{L}(K^{(\ell)}+1)}, \sum_{\ell=1}^{L}\sum_{j=0}^{K^{(\ell)}} x_j^{(\ell)} = 1\}.
\tag{5}$$

The four possible changes in the number of queues with length i which appear in (4) are:

- A job arrives to a queue with length $i - 1$;
- A job arrives to a queue with length i;
- A job leaves a queue with length $i + 1$;

– A job leaves a queue with length i.

On the border of the domain (5), certain changes cannot occur. There is no service in empty queues or arrival to full queues:

$$\mu_0^{(\ell)} = 0 \quad (\ell = 1, \ldots, L), \qquad f_{K^{(\ell)}}(.) \equiv 0 \quad (\ell = 1, \ldots, L).$$

The general transient mean-field equation corresponding to (4) is

$$
\begin{aligned}
v_i^{(\ell)}(t) = v_i^{(\ell)}(0) &+ \int_0^t \lambda f_{i-1}^{(\ell)}(v(s)) \mathrm{d}s - \int_0^t \lambda f_i^{(\ell)}(v(s)) \mathrm{d}s \\
&+ \int_0^t \mu_{i+1}^{(\ell)} v_{i+1}^{(\ell)}(s) \mathrm{d}s - \int_0^t \mu_i^{(\ell)} v_i^{(\ell)}(s) \mathrm{d}s.
\end{aligned}
\tag{6}
$$

in integral form, or

$$\frac{\mathrm{d}}{\mathrm{d}t} v_i^{(\ell)}(t) = \lambda f_{i-1}^{(\ell)}(v(t)) - \lambda f_i^{(\ell)}(v(t)) + \mu_{i+1}^{(\ell)} v_{i+1}^{(\ell)}(t) - \mu_i^{(\ell)} v_i^{(\ell)}(t) \tag{7}$$

in differential form. An empty initial cluster corresponds to the initial condition

$$v_i^{(\ell)}(0) = \begin{cases} \gamma_\ell & \text{for } i = 0, \\ 0 & \text{otherwise.} \end{cases}$$

Theorem 1 guarantees transient mean-field convergence whenever the $f_i^{(\ell)}$ functions are Lipschitz-continuous. Although we do not chase rigorous proofs in the present paper, we conjecture that the conclusions of Theorem 1 still apply for all the discontinuous choices of $f_i^{(\ell)}$ in the present paper.

In the continuous case, the equations for the mean-field stationary distribution ν can be obtained from (7) by setting $\frac{\mathrm{d}}{\mathrm{d}t} v_i^{(k)}(t) = 0$:

$$
\begin{aligned}
0 = \lambda f_{i-1}^{(\ell)}(v(t)) &- \lambda f_i^{(\ell)}(v(t)) + \mu_{i+1}^{(\ell)} v_{i+1}^{(\ell)}(t) - \mu_i^{(\ell)} v_i^{(\ell)}(t) \\
&i \in \{1, \ldots, K^{(\ell)-1}\}, \qquad \ell \in \{1, \ldots, L\}
\end{aligned}
\tag{8}
$$

which are equivalent to the dynamic balance equations

$$\mu_i^{(\ell)} v_i^{(\ell)} = \lambda f_{i-1}^{(\ell)}(\nu), \qquad i \in \{1, \ldots, K^{(\ell)}\}, \qquad \ell \in \{1, \ldots, L\}. \tag{9}$$

We also have equations for the ratio of each server type

$$\sum_{i=0}^{B^{(\ell)}} \nu_i^{(\ell)} = \gamma_\ell \qquad \ell \in \{1, \ldots, L\}. \tag{10}$$

(9) + (10) provide algebraic equations for ν. Another way to compute ν numerically is to numerically solve the transient Equations (7) and take the solution at a large enough time. (This assumes convergence to a single stationary solution, which we do not aim to prove rigorously either.)

For systems with a finite number of servers, the global state might fluctuate due to arrival and service in other queues, but the fluctuations go to 0 as the system size increases, vanishing in the mean-field limit. With that in mind, the mean-field limit process can be regarded as a queueing process with an infinite (or extremely large) number of servers, where the mean-field stationary distribution ν acts as a stationary environment for a tagged job. With this interpretation, tools for finite queueing systems such as Little's Law also apply to the mean-field limit, and will allow us to compute performance metrics such as the mean system time of the tagged job (see Sect. 3).

When the $f_i^{(\ell)}$ are discontinuous at ν, (9) no longer applies and needs to be modified; the modified balance equations will be derived in Sect. 3.

As for stationary mean-field convergence, we conjecture the conclusions of Theorem 1 apply for all (continuous and discontinuous) choices of $f_i^{(\ell)}$ in the present paper, with the computation of ν different for the continuous and discontinuous cases.

3 General Prioritization Scheme

With Join-Shortest-Queue (JSQ) load balancing policy, queues are ordered according to queue length, and upon a job arrival, the job is dispatched to the shortest queue (or one of the shortest queues at random if there are multiple). Other load balancing policies may use a different prioritization scheme; in this section, we present a general prioritization scheme that also considers server types.

Assume that all (ℓ, i) pairs $(0 \leq i \leq K^{(\ell)-1}, \ell = 1, \ldots, L)$ are fully ordered (with ties allowed). This can be implemented using a prioritization function $V(\ell, i)$: the incoming job is dispatched to a queue with minimal V value, or, in case multiple queues are tied for minimal V, to one of the tied queues at random. The actual values of the function V are not relevant, only their relative order (including ties).

For classic JSQ, the prioritization function $V(\ell, i)$ is simply the queue length i. While this is straightforward for FIFO queues, for M/M/c queues with $c > 1$, there is some ambiguity; it also makes sense to consider the prioritization function

$$V(\ell, i) = \begin{cases} 0 & \text{if } i < c^{(\ell)}, \\ i - c^{(\ell)} + 1 & \text{if } i \geq c^{(\ell)} \end{cases} \tag{11}$$

since an incoming job can start service either immediately if $i < c^{(\ell)}$, or after $i - c^{(\ell)} + 1$ jobs are served for $i \geq c^{(\ell)}$. The prioritization function (11) will be referred to as JSQ-A, and the prioritization function $V(\ell, i) = i$ will be referred to as JSQ-B for distinction.

Other classic load balancing principles also fit in this framework:

– JIQ (join-idle-queue) load balancing principle corresponds to $V(\ell, i) = 1\{i > 0\}$.

- JBT (join-below-threshold) is obtained when V is set to 0 for queues below the threshold and 1 for queues at or above the threshold (the threshold may depend on the queue type).
- a constant V corresponds to random dispatch (no load balancing).

Since the function V can depend on the queue type ℓ, type-aware variants of JSQ also fit in this framework; we define one such load balancing principle:

- JPSQ (join-proportionally-shortest-queue): V is set to the mean system time to finish, depending on the queue type and length.

JPSQ utilizes the fact that for M/M/c and M/M/c/K servers, the system time distribution of a job does not depend on jobs arriving later, so the mean system time (depending on the queue type and length) is straightforward to compute and can be used to choose the queue to enter. (In fact, this is the main motivation in choosing M/M/c/K servers.)

JPSQ corresponds to an individual optimal choice for each job. However, this does not directly imply that the mean system time of jobs is minimal globally, since the choice of load balancing principle also affects the mean-field stationary distribution ν. For the numerical examples in Sect. 5, JPSQ turns out to be optimal, but the numerical analysis is far from exhaustive, so we refrain from conjecturing JPSQ as always optimal, just mention it as a strong candidate for the optimum.

In the literature, JPSQ is also known as shortest expected delay (SED), which has been examined in e.g. [4], albeit only for FIFO servers, for which the structure of the mean field stationary distribution is almost trivial (only concentrated on queues of length 0 or 1).

This section focuses on the mean-field analysis for a general prioritization function; specifically, the computation of the mean-field stationary distribution ν and the mean system time of jobs W.

To demonstrate each step of the method, we will have a running example.

Running example. Consider a system with $L = 2$ server types with parameters according to Table 1. Type 1 servers are FIFO queues, while type 2 servers can serve up to 5 jobs simultaneously.

Table 1. Running example - system parameters ($L = 2$ server types)

ℓ	$\gamma^{(\ell)}$	$K^{(\ell)}$	$c^{(\ell)}$	$\mu^{(\ell)}$	λ
1	0.75	6	1	1	1.6
2	0.25	6	5	0.8	

The system has JPSQ load balancing principle, which corresponds to the V function (in matrix form, with the rows indexed $\ell = 1, 2$ and the columns indexed $i = 0, 1, \ldots, 5$)

$$V = \begin{bmatrix} 1 & 2 & 3 & 4 & 5 & 6 \\ 1.25 & 1.25 & 1.25 & 1.25 & 1.25 & 1.5 \end{bmatrix}. \tag{12}$$

The values in the first row simply come from the FIFO queue with service rate 1; a job that is at position i in the queue will have mean system time i. The second row corresponds to the mean system time of jobs entering type 2 servers, where up to 5 jobs can be served the first row corresponds to the mean system time of a job entering a server of type 2; as long as the server has fewer than 5 jobs, it has a free thread that can start service with rate 0.8 immediately, giving a mean system time of 1.25. The bottom right element of (12) corresponds to the mean system time of a job in a type 2 server that has to wait until one of the 5 threads finishes, which adds a mean time of $1/(5 \times 0.8) = 0.25$ compared to the other elements of the second row.

We note that for the function V, only the relative order of the elements is relevant (including ties), so any other table with the same relative order corresponds to the same load balancing principle.

For a given $V(\ell, i)$ function, the dispatch functions in the Poisson representation (4) are

$$f_i^{(\ell)}(x) = \begin{cases} 0 & \text{if } (\ell, i) \notin T(x), \\ \dfrac{x_i^{(\ell)}}{\sum\limits_{(\ell, i) \in T(x)} x_i^{(\ell)}} & \text{if } (\ell, i) \in T(x), \end{cases} \tag{13}$$

where

$$V_0(x) = \min\{V(\ell, i) : x_i^{(\ell)} > 0\}, \qquad T(x) = \{(\ell, i) : V(\ell, i) = V_0(x)\}. \tag{14}$$

$T(x)$ denotes the set of (ℓ, i) pairs tied for minimal $V(\ell, i)$ in a given state x.

The dispatch functions (13) are discontinuous at any point x with $\sum_{\ell=1}^{L} x_0^{(\ell)} = 0$, which means that as long as ν falls within this region, (8) and (9) cannot be used to compute the mean-field stationary distribution ν. We address the nature of this discontinuity and provide the modified version of (9) to compute ν next.

Stationary Mean-Field Setup. In the same spirit as the definition of $V_0(x)$ in (14), for the mean-field stationary distribution ν, there will be a critical value V_0 which we aim to define next. The value of V_0 depends on the arrival rate λ, but in subsequent formulas, the dependence on λ is not denoted.

For a given $V(\ell, i)$ function and λ arrival rate, we define the following quantities:

$$\hat{\imath}^{(\ell)}(V) := \operatorname{argmax}\{i : V_i^{(\ell)} \leq V\}, \tag{15}$$

with $\hat{\imath}^{(\ell)}(V)$ defaulting to 0 if the right hand side is empty. Then

$$V_0 := \operatorname{argmin}\left\{ V : \sum_{\ell=1}^{L} \gamma_\ell \mu_{\hat{\imath}^{(\ell)}(V)}^{(\ell)} \geq \lambda \right\}.$$

Basically, V_0 corresponds to the critical point in the prioritization ordering where the total service capacity reaches λ. We elaborate a little more. In ν, if all queues are filled up to a value $V < V_0$ (i.e. for each queue of type ℓ, its queue length i is such that $V(\ell, i) \leq V$), then the total service capacity of the entire system will be below λ (keeping in mind that the service rate curve is increasing with the queue length (2)). That is not possible, since the mean-field stationary distribution is a dynamic balance where the total service rate must be equal to the total arrival rate. Accordingly, in ν, all (ℓ, i) pairs with a positive weight must satisfy $V(\ell, i) \geq V_0$.

Running Example. In the running example, $V_0 = 1.25$.

In ν, there may be (ℓ, i) pairs with either $V(\ell, i) = V_0$ or $V(\ell, i) > V_0$. Arrivals will never be dispatched to queues with $V(\ell, i) > V_0$; such queues may only result from arrivals to queues with value $V(\ell, i) \leq V_0$. Queues with value $V(\ell, i) < V_0$ do not appear with a positive weight in ν, but in the dynamic balance, service in queues with value $V(\ell, i) \geq V_0$ may still result in a queue with value $V(\ell, i) < V_0$. However, since such a queue has a lower value than all other queues, if such a queue appears, all arrivals will be dispatched to this single queue; in the mean-field limit, this means that such a queue will be filled back up to value $V(\ell, i) \geq V_0$ immediately (and will not be visible for any positive amount of time).

In other words, in the mean-field stationary regime, service instances which result in a queue with value $V(\ell, i) < V_0$ will be balanced out by an arrival immediately. Altogether, a positive proportion of arrivals will be dispatched this way. Such arrivals will be collectively referred to as *the upkeep*. Actually, the presence of the upkeep corresponds directly to the discontinuous dispatch functions; in systems where the dispatch functions are continuous, there is no upkeep.

Deriving the Modified Balance Equations. We aim to modify the balance equations (9) to include the upkeep. We define

$$\mathfrak{i}^{(\ell)} := \operatorname{argmin}\{i : V(\ell, i) \geq V_0\}. \tag{16}$$

The argmin in (16) is never empty, so $\mathfrak{i}^{(\ell)}$ is well defined.

In ν, for each queue type ℓ, there are no queues shorter than $\mathfrak{i}^{(\ell)}$ with positive weight. Queues of length $\mathfrak{i}^{(\ell)}$ contribute to the upkeep; the total service rate in such queues is

$$\lambda_u = \sum_{\ell=1}^{L} \mu_{\mathfrak{i}(\ell)}^{(\ell)} \nu_{\mathfrak{i}(\ell)}^{(\ell)}. \tag{17}$$

We note here that $\mathfrak{i}^{(\ell)} = 0$ is possible, and corresponds to a queue type ℓ which do not contribute to the upkeep. Accordingly, we set $\mu_0^{(\ell)} = 0$ in (17) if necessary. The value λ_u will also be referred to as the upkeep.

We have $\lambda_u \leq \lambda$ due to the definition of V_0 (15). Overall, out of the total incoming rate λ, λ_u will be dispatched according to the upkeep, and the remaining $\lambda - \lambda_u$ will be dispatched among queues with value tied for V_0. We will refer to this part of the arrivals as *dynamic dispatch*.

Next we look to identify the quantities relevant for dynamic dispatch. Define

$$\hat{\imath}^{(\ell)} := \operatorname{argmax}\{i : V(\ell, i) \leq V_0\}. \tag{18}$$

In (18), the argmax may be empty; in this case, we set $\hat{\imath}^{(\ell)} := \check{\imath}^{(\ell)} - 1$. For a given ℓ, $\hat{\imath}^{(\ell)} = \check{\imath}^{(\ell)} - 1$ also holds in case $V(\ell, i) \neq V_0$ for any queue length i.

For a given queue type ℓ, the interval $[\check{\imath}^{(\ell)}, \ldots, \hat{\imath}^{(\ell)}]$ contains all queue lengths i for which $V(\ell, i) = V_0$. Depending on the number of queue lengths i for which $V(\ell, i) = V_0$ holds, this interval may be empty, or it may consist of a single queue length i, or it may consist of several different queue lengths i.

For any queue type ℓ, ν may assign positive weight only to pairs (ℓ, i) with $\check{\imath}^{(\ell)} \leq i \leq \hat{\imath}^{(\ell)} + 1$. Dynamic dispatch targets only queues with $\check{\imath}^{(\ell)} \leq i \leq \hat{\imath}^{(\ell)}$; queues of length $\hat{\imath}^{(\ell)} + 1$ may only result from arrivals to queues of length $\hat{\imath}^{(\ell)}$.

Running example. In the running example, from $V_0 = 1.25$ it follows that $\check{\imath}^{(1)} = 1, \hat{\imath}^{(1)} = 0$, which means that dynamic dispatch does not target servers of type 1, and $\check{\imath}^{(2)} = 0, \hat{\imath}^{(2)} = 4$, so dynamic dispatch is distributed among queues of type 2 with queue length $0, \ldots, 4$.

Now we are ready to derive the balance equations in general. The dynamic dispatch will target queues of length $\check{\imath}^{(\ell)}, \ldots, \hat{\imath}^{(\ell)}$ for each $\ell = 1, \ldots, L$ at random. The total weight of these queues in ν is

$$d_0 := \sum_{\ell=1}^{L} \sum_{i=\check{\imath}^{(\ell)}}^{\hat{\imath}^{(\ell)}} \nu_i^{(\ell)}, \tag{19}$$

and accordingly, the dynamic balance equation between queues of type ℓ and length i and $i + 1$ is

$$\frac{\lambda - \lambda_u}{d_0} \nu_i^{(\ell)} = \mu_{i+1}^{(\ell)} \nu_{i+1}^{(\ell)} \tag{20}$$

$$(\check{\imath}^{(\ell)} \leq i \leq \hat{\imath}^{(\ell)}, \quad \ell = 1, \ldots, L),$$

where λ_u comes from (17), and we also have the corresponding version of (10), which can now be written as

$$\sum_{i=\check{\imath}^{(\ell)}}^{\hat{\imath}^{(\ell)}+1} \nu_i^{(\ell)} = \gamma_\ell \qquad \ell \in \{1, \ldots, L\}. \tag{21}$$

Solving the Balance Equations. Next we solve the system (20)+(17)+(21), based on a similar calculation in [2]. We utilize the fact that in (20), the term

$$x := \frac{\lambda - \lambda_u}{d_0} \tag{22}$$

is unknown, but constant throughout the equations. Putting x in (20) yields

$$\nu_{i+1}^{(\ell)} = \frac{x}{\mu_{i+1}^{(\ell)}} \nu_i^{(\ell)} \qquad (\check{\imath}^{(\ell)} \leq i \leq \hat{\imath}^{(\ell)}),$$

which in turn gives

$$\nu_i^{(\ell)} = \nu_{\check{\imath}^{(\ell)}}^{(\ell)} \prod_{j=\check{\imath}^{(\ell)}+1}^{i} \frac{x}{\mu_j^{(\ell)}} \qquad (\check{\imath}^{(\ell)} \leq i \leq \hat{\imath}^{(\ell)} + 1), \tag{23}$$

which can be put into (21) to obtain

$$\gamma_l = \nu_{\check{\imath}^{(\ell)}}^{(\ell)} \sum_{i=\check{\imath}^{(\ell)}}^{\hat{\imath}^{(\ell)}+1} \prod_{j=\check{\imath}^{(\ell)}+1}^{i} \frac{x}{\mu_j^{(\ell)}} \qquad \ell \in \{1, \ldots, L\}$$

(with the empty product for $i = \check{\imath}^{(\ell)}$ defaulting to 1). Then

$$\nu_{\check{\imath}^{(\ell)}}^{(\ell)} = \frac{\gamma_l}{\sum_{i=\check{\imath}^{(\ell)}}^{\hat{\imath}^{(\ell)}+1} \prod_{j=\check{\imath}^{(\ell)}}^{i-1} \frac{x}{\mu_{j+1}^{(\ell)}}} \qquad \ell \in \{1, \ldots, L\}, \tag{24}$$

which can then be put into (17):

$$\lambda_u = \sum_{\ell=1}^{L} \frac{\gamma_l \mu_{\check{\imath}^{(\ell)}}^{(\ell)}}{\sum_{i=\check{\imath}^{(\ell)}}^{\hat{\imath}^{(\ell)}+1} \prod_{j=\check{\imath}^{(\ell)}}^{i-1} \frac{x}{\mu_{j+1}^{(\ell)}}}.$$

At this point, all $\nu_i^{(\ell)}$'s are expressed as a function of x; we put them all into the definition of d_0 (19) to obtain

$$d_0 = \sum_{\ell=1}^{L} \gamma_l \frac{\sum_{i=\check{\imath}^{(\ell)}}^{\hat{\imath}^{(\ell)}} \prod_{j=\check{\imath}^{(\ell)}}^{i-1} \frac{x}{\mu_{j+1}^{(\ell)}}}{\sum_{i=\check{\imath}^{(\ell)}}^{\hat{\imath}^{(\ell)}+1} \prod_{j=\check{\imath}^{(\ell)}}^{i-1} \frac{x}{\mu_{j+1}^{(\ell)}}}, \tag{25}$$

$$x = \left(\lambda - \sum_{\ell=1}^{L} \frac{\gamma_l \mu_{\check{\imath}^{(\ell)}}^{(\ell)}}{\sum_{i=\check{\imath}^{(\ell)}}^{\hat{\imath}^{(\ell)}+1} \prod_{j=\check{\imath}^{(\ell)}}^{i-1} \frac{x}{\mu_{j+1}^{(\ell)}}}\right) \Bigg/ \left(\sum_{\ell=1}^{L} \gamma_l \frac{\sum_{i=\check{\imath}^{(\ell)}}^{\hat{\imath}^{(\ell)}} \prod_{j=\check{\imath}^{(\ell)}}^{i-1} \frac{x}{\mu_{j+1}^{(\ell)}}}{\sum_{i=\check{\imath}^{(\ell)}}^{\hat{\imath}^{(\ell)}+1} \prod_{j=\check{\imath}^{(\ell)}}^{i-1} \frac{x}{\mu_{j+1}^{(\ell)}}}\right), \tag{26}$$

which reduces to a polynomial equation for x. Numerical experiments show that this equation only has one feasible solution for x. Without providing a rigorous argument, we just note that a monotonicity based argument might be an option, again, similar to [2].

Once x is obtained from (26), the $\nu_i^{(\ell)}$ values follow from (24) and (23).

For M/M/c/K queues, there are several ways to compute the mean system time of jobs from ν. One approach is to note that the mean system time of a job depends only on the length of its queue upon entering the system. Another approach is via Little's Law: from ν, compute the effective arrival rate

$$\lambda_e = \lambda_u + (\lambda - \lambda_u)\frac{\sum_{\ell=1}^{L}\sum_{i=0}^{K^{(\ell)}-1} f_i^{(\ell)}(\nu)}{\sum_{\ell=1}^{L}\sum_{i=0}^{K^{(\ell)}} f_i^{(\ell)}(\nu)}.$$

λ_e excludes the rate of jobs lost due to being dispatched to a full queue. Job loss is never possible for the upkeep, hence the full λ_u rate contributes to λ_e; from the dynamic dispatch part, jobs dispatched to full queues have to be excluded.

The average queue length is

$$A = \sum_{\ell=1}^{L}\sum_{i=1}^{K^{(\ell)}} i\nu_i^{(\ell)},$$

and, according to Little's Law, the mean system time of a job is

$$W = A/\lambda_e.$$

4   Optimization Procedure for the Load Balancing Principle

In accordance with Sect. 2, a server cluster is described by the following set of parameters:

- the number of server types L;
- the ratios of each server type: $\gamma_1, \ldots, \gamma_L$;
- the buffer sizes $K^{(1)}, \ldots, K^{(L)}$;
- the number of threads $c^{(1)}, \ldots, c^{(L)}$;
- the service rates of single threads $\mu^{(1)}, \ldots, \mu^{(\ell)}$;
- the arrival rate λ.

We aim to find the optimal load balancing principle for the cluster, where optimal corresponds to the minimal mean system time of jobs. For a given set of parameters, there is only a finite number of possible prioritization functions V (keeping in mind that only the relative order of the values of V is relevant, so we can set the values of the function e.g. as positive integers starting from 1), so this is actually a discrete optimization problem, where the naive approach would be to compute the mean system time W for all possible choice of V and select the optimum.

The naive algorithm can be improved by reducing the number of prioritization function included in the search based on several different ideas. Some possibilities are the following:

- Prioritization functions which allow data loss in the mean-field limit can be identified in advance and excluded from the optimization. (These may include prioritization functions not using certain queue types at all.)

- For $i > \hat{\imath}^{(\ell)}$, jobs will never be dispatched to queues of length i and type ℓ, thus the exact prioritization order among such (ℓ, i) pairs is irrelevant in the stationary mean-field limit. (In other words, exact prioritization order matters only up to $\hat{\imath}^{(\ell)}$ for each $\ell = 1, \ldots, L$.) Prioritization functions that only differ for (ℓ, i) pairs with $i > \hat{\imath}^{(\ell)}$ can be considered equivalent and it is sufficient to include one representative function from among each equivalency class.
- It may be possible to exclude some suboptimal prioritizations in advance based on other heuristics; this is subject to further research.

Currently, the naive approach is implemented with the first idea above added, so the optimization method is the following: list all prioritization functions, then for each choice of V (excluding prioritization functions with data loss), compute the mean-field stationary distribution ν and the mean system time W according to Sect. 3, then select the minimal W.

The optimizer code is available online in python, see [1].

5   Numerical Results

5.1   Mean-Field and Stationary Convergence

In this section, we examine mean-field convergence and convergence to stationarity; this is not directly related to the optimization, just serves to showcase the various notions of convergence present in the paper. We consider a simple homogeneous system with parameters according to Table 2.

Table 2. System parameters ($L = 1$ server type)

ℓ	$\gamma^{(\ell)}$	$K^{(\ell)}$	$c^{(\ell)}$	$\mu^{(\ell)}$	λ
1	1	5	5	1	2.5

The considered load balancing principle is JSQ (JSQ-B with the notation of Sect. 3).

Figure 1 displays the transient evolution of two such systems: with $N = 500$ servers on the left, and with $N = 10000$ servers on the right.

Both systems start from an empty state at time $t = 0$. The jagged lines correspond to the global state of the system, that is, the ratio of servers with 0 to 3 jobs in their queue respectively. The smooth thin curves correspond to the transient mean-field limit of the system (see Theorem 1 and also (7)). For the finite systems, a discrete event simulator was used, while for the transient mean-field limit, a numerical differential inclusion solver was used. All codes are available in python at [1].

For both $N = 500$ and $N = 10000$, the following behaviour can be observed in Fig. 1:

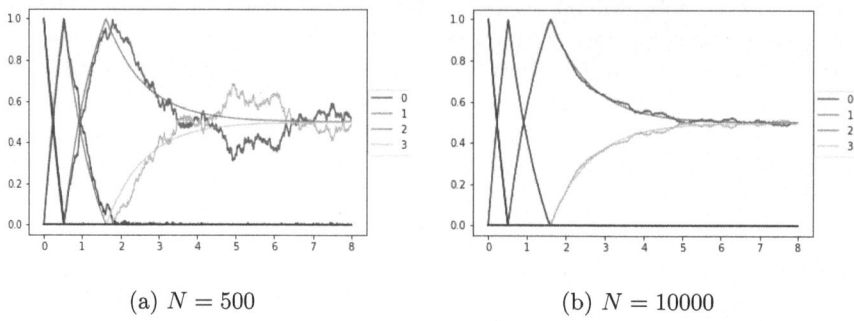

(a) $N = 500$ \qquad\qquad (b) $N = 10000$

Fig. 1. Transient system behaviour for various choices of N

- The system starts from an empty state, where, due to JSQ, all arrivals are directed to empty queues. Since $\lambda > \mu$, the queues will start to fill up to length 1, until they reach a point with no more empty queues (leftmost breaking point in the curves).
- After that, queues with length 1 will begin to fill up to length 2 due to $\lambda > 2\mu$, until they reach a point with no more queues of length 1 (second breaking point from the left).
- After that, queues of length 2 will begin to fill up to length 3, but, since $2\mu < \lambda < 3\mu$, the servers now stabilize at a mixture of queue lengths 2 and 3. Since λ is actually halfway between 2μ and 3μ, dynamic balance is reached at $\nu_2 = \nu_3 = 1/2$, which is actually the mean-field stationary distribution of this system.
- As time goes to infinity, the mean-field system (smooth curves) converges to the mean-field stationary distribution.
- As time goes to infinity, for any fixed N, the finite system stabilizes near the mean-field limit, but with fluctuations that do not vanish even as time is increased.
- For larger N, the fluctuations around the transient mean-field limit are smaller, and go to 0 as $N \to \infty$.

5.2   Mean System Time Optimization

In this section, we first examine the running example, with parameters according to Table 1.

The optimization algorithm of Sect. 4 gave the V function

$$V = \begin{bmatrix} 1\ 3\ 4\ 5\ 6\ 7 \\ 2\ 2\ 2\ 2\ 2\ 7 \end{bmatrix}, \tag{27}$$

and the optimal mean system time $W = 1.1328$.

We note that (27) is actually equivalent to the prioritization considered in the running example (12) since the elements of the two matrix have the same relative ordering. This means that for the running example, the optimal load balancing is actually JPSQ (which differs from both JSQ-A and JSQ-B, which are thus non-optimal).

We consider one more example with $L = 2$ server types with parameters according to Table 3. Both server types have 3 threads, but servers of type 2 are slower.

Table 3. System parameters ($L = 2$ server types)

ℓ	$\gamma^{(\ell)}$	$K^{(\ell)}$	$c^{(\ell)}$	$\mu^{(\ell)}$	λ
1	0.5	6	3	1	2.0
2	0.5	6	3	0.5	

Optimization gives

$$V = \begin{bmatrix} 1\,1\,1\,2\,3\,4 \\ 2\,2\,2\,4\,5\,6 \end{bmatrix}, \tag{28}$$

and the optimal mean system time $W = 1.25$. Once again, this turns out to be equivalent to JPSQ since JPSQ prioritizes servers of type 1 as long as there is at least 1 free thread; for an arrival rate of $\lambda = 2.0$, servers of type 1 will be in use 100% of the time, with total service rate $0.5 \times 3 \times 1 = 1.5$, while the remaining $2.0 - 1.5 = 0.5$ arrival rate will be served by servers of type 2. Overall, $1.5/2.0 = 75\%$ of jobs will be served by servers of type 1 with mean system time 1, while $0.5/2.0 = 25\%$ of jobs will be served by servers of type 2 with mean system time $1/0.5 = 2$, and the aggregate mean system time is indeed

$$0.75 \times 1 + 0.25 \times 2 = 1.25.$$

We also note that according to the second improvement idea of Sect. 4, there are many equivalent versions of essentially the same prioritization; one simple way to identify them collectively is via the notation

$$V = \begin{bmatrix} 1\,1\,1\,2 * * \\ 2\,2\,2 * * * \end{bmatrix},$$

where the exact ordering of the * elements is not relevant as long as they are all larger than 2.

5.3   Mean-Field Approximation of the Mean System Time

The main contribution of the paper is a method for finding the prioritization function (and corresponding load balancing method) that minimizes mean system time in the mean-field limit. Next we examine how the mean system time in the mean-field limit compares to the mean system time in finite versions of the same system (of various sizes). No rigorous results (e.g. convergence rates) were given in the present paper; instead, we demonstrate the rate of convergence numerically.

We pick the same heterogeneous system as in Table 1 from Sect. 5.2 with JPSQ load balancing, so the mean system time in the mean-field regime is $W = 1.1328$. This is compared with the mean system time of several finite systems with the same parameters for various choices of N.

So for example for $N = 20$, $\gamma_1 = 0.75$ and $\gamma_2 = 0.25$ means that there are $N_1 = \gamma_1 N = 15$ and $N_2 = \gamma_2 N = 5$ servers of the two types. The rest of the parameters are in accordance with Table 1.

Table 4 shows the mean system time for various different choices of a finite N, and also the mean-field limit ($N = \infty$). For finite values of N, a discrete event simulator was used for (4), with an initially empty system. The time to reach stationarity can be estimated based on the transient mean-field behaviour (see also Fig. 1 in Sect. 5). For the mean system time statistics in Table 4, only the stationary regime was considered. Enough independent runs were averaged so that the standard deviation of each value in Table 4 is < 0.001 (except for the mean system time for $N = \infty$, which is not random).

Table 4. Mean system time

	$N = 20$	$N = 100$	$N = 500$	$N = \infty$
mean system time	1.3370	1.1483	1.1350	1.1328

Table 4 shows that the mean system time does seem to converge to the mean-field limit as $N \to \infty$, with the actual value very close already for $N = 100$. For smaller values of N, the mean system time in the stationary mean-field limit can still serve as an approximation with reasonably some error.

The convergence seems to hold in a descending manner, that is, the mean system time is larger for smaller systems. A possible intuitive explanation is that smaller systems typically have larger fluctuations, creating congestions; while temporary, these have a relatively high number of jobs with increased system time, and occur frequently enough to have a significant effect on the mean system time.

6   Conclusion

We presented a mathematical framework that allows the efficient computation of the mean system time of jobs in the stationary mean-field limit for heterogeneous M/M/c/K clusters with any prioritization based on queue type and length, then presented a discrete optimization method to find the optimal prioritization.

Depending on the parameters (notably for larger values of L and $K^{(1)}, \ldots, K^{(L)}$), it might be necessary to reduce the number of prioritization functions based on some intuitions to make the runtime feasible.

Different research directions include prioritization schemes where the prioritization may depend on factors other than the queue type and queue length (e.g. physical locations). In such settings, we expect the mean-field limit to be different from what the results of Kurtz dictate.

The framework can be modified to include other factors (e.g. overhead computational costs of load balancing and server-dispatcher communication). However, these would be highly scenario-dependent, and should only be addressed with specific real-life scenarios in mind.

Acknowledgments. I. Horváth was supported by the OTKA K-138208 project of the Hungarian Scientific Research Fund.

References

1. Load balancing mean-field optimizer. https://github.com/mezsimarci/lbmfcodes. Accessed 24 Apr 2024
2. Antal Horváth, I., Scully, Z., Van Houdt, B.: Mean field analysis of join-below-threshold load balancing for resource sharing servers. SIGMETRICS Perform. Eval. Rev. **48**(1), 41–42 (2020)
3. Benaïm, M., Le Boudec, J.-Y.: A class of mean field interaction models for computer and communication systems. Performance Eval. **65**(11), 823–838 (2008). Performance Evaluation Methodologies and Tools: Selected Papers from ValueTools 2007
4. Bhambay, S., Mukhopadhyay, A.: Asymptotic optimality of speed-aware jsq for heterogeneous service systems. Perform. Eval. **157–158**, 102320 (2022)
5. Ethier, S.N., Kurtz, T.G.: Markov Processes: Characterization and Convergence. Wiley (2005)
6. Gast, N., Gaujal, B.: Markov chains with discontinuous drifts have differential inclusion limits. Performance Eval. **69**, 623–642 (2012)
7. Horvath, I., Meszaros,, M.: Mean-field analysis of load balancing principles in large scale systems. https://arxiv.org/abs/2307.04360 (2023)
8. Kurtz, T.: Solutions of ordinary differential equations as limits of pure jump Markov processes. J. Appl. Probability **7**, 49–58 (1970)
9. Kurtz, T.G.: Strong approximation theorems for density dependent Markov chains. Stochastic Processes Appl. **6**(3), 223–240 (1978)
10. Lin, H.-C., Raghavendra, C.S.: Approximating the mean response time of parallel queues with JSQ policy. Comput. Oper. Res. **23**(8), 733–740 (1996)
11. Mitzenmacher, M.: The power of two choices in randomized load balancing. IEEE Trans. Parallel Distrib. Syst. **12**, 1094–1104 (2001)
12. Mitzenmacher, M.: Analyzing distributed join-idle-queue: a fluid limit approach. In *2016 54th Annual Allerton Conference on Communication, Control, and Computing (Allerton)*, pp. 312–318, September 2016
13. Roth, G., Sandholm, W.H.: Stochastic approximations with constant step size and differential inclusions. SIAM J. Control. Optim. **51**(1), 525–555 (2013)
14. Tsitsiklis, J.N., Xu, K.: On the power of (even a little) centralization in distributed processing. In: Proceedings of the ACM SIGMETRICS Joint International Conference on Measurement and Modeling of Computer Systems, SIGMETRICS 2011, pp. 161–172. Association for Computing Machinery, New York (2011)

An Algebraic Proof of the Relation of Markov Fluid Queues and QBD Processes

Peter Buchholz[1] ⓘ, Andras Meszaros[2,3] ⓘ, and Miklos Telek[2,3(✉)] ⓘ

[1] Technische Universität Dortmund, Dortmund, Germany
peter.buchholz@udo.edu
[2] Department of Networked Systems and Services, Budapest University of Technology and Economics, Budapest, Hungary
[3] HUNREN-BME Information Systems Research Group, Budapest, Hungary
{meszarosa,telek}@hit.bme.hu

Abstract. We study the relation of Markov fluid queues and QBD processes in this paper. Ahn and Ramaswami presented results about this relation and provided a stochastic interpretation based reasoning in [1]. In the current work, first we provide an algebraic proof for that relation.

After that, we present a negative result about the potential extension of the QBD based analysis Markov fluid queues to Markov fluid queues with two buffers. We present a 2-dimensional QBD process, which could be a candidate for describing the stationary behaviour of the related Markov fluid queue, but it turns out that the QBD based behaviour is different from the one of the Markov fluid queue.

Keywords: Quasi birth death process · Markov fluid queue · queue with 2 buffers

1 Introduction

Markov fluid queues MFQs have a long tradition in stochastic modeling [5]. MFQs are queueing models where the queue length is continuous and the rate at which the queue length (also referred to as fluid level) changes is modulated by a background continuous time Markov chain (CTMC). Since the seminal applications of MFQs for analyzing high speed networks in [4], they have been applied successfully in many application areas, e.g., [13,18].

Several solution methods have been developed to obtain the stationary distribution of the fluid level in MFQs (e.g., eigenvalue decomposition based [14], Schur decomposition based [3], matrix-analytic [12,17], invariant subspace based [2], etc.).

In [1], Ahn and Ramaswami provide stochastic interpretation based results about the relation of MFQs and quasi birth death (QBD) processes. The relation is based on a stochastic coupling argument and shows a correspondence between

© The Author(s), under exclusive license to Springer Nature Switzerland AG 2025
A. Devos et al. (Eds.): ASMTA 2024, LNCS 14826, pp. 132–147, 2025.
https://doi.org/10.1007/978-3-031-70753-7_9

the fluid level and the virtual workload in a queue. In this work we present an algebraic proof for the relation between MFQs and QBDs.

In [1] the authors mention that more than one QBD structure can establish the relation with the fluid queue, but they consider only one such QBD structure. In the current work we consider another QBD structure, when we prove the relation of fluid queues and QBD models.

The main advantage of the QBD interpretation of MFQs is the possibility of using efficient methods for QBD [15] analysis also for the analysis of MFQs. Thus, it is natural to ask whether the relation between MFQs and QBDs also can be established for fluid models with more than one buffer like tandem systems [16], fluid-fluid systems [7] or parallel buffers filled by one source but emptied with different rates [10]. Although there are some first results on the stationary analysis of such models [6,8,11], the proposed methods are often computationally expensive and numerically unstable. Thus, a relation to QBDs or level dependent QBDs [9] would be an important step towards an efficient analysis of more general fluid models.

In this paper we introduce two different level dependent QBDs to model fluid models with two parallel buffers. Unfortunately, as we show, the partial differential equations (PDEs) describing the fluid flows in the buffers differ from the PDEs of the original systems. Consequently, the simple relation between QBDs and MFQs that holds for one buffer, no longer holds for two buffers. This is a negative result which, nevertheless, helps to understand the behavior of more complex fluid queues.

The rest of the paper is organized as follows. In the following section we introduce the later used properties of MFQs and QBDs processes. In Sect. 3 we discuss the relation of MFQs and QBD processes. In Sect. 4, a queuing model with two fluid buffers is introduced together with its stationary equations, and after that we study a level dependent quasi birth death (LDQBD) process, whose behaviour mimics the one of the MFQ with two fluid buffers. As a result of this analysis we show that the behavior of the two models differ. The paper is finally concluded in Sect. 5.

2   Basic Properties of Markov Fluid Queues and Quasi Birth Deaths Processes

2.1   Markov Fluid Queues

We consider an infinite buffer MFQ $(Y(t), J(t))$, where $J(t) \in \mathcal{S}$ is the state of the background CTMC and $Y(t)$ is the fluid level at time t. We assume non-zero fluid rates, such that the rates are positive when the CTMC visits a state in $\mathcal{S}_+$ and negative when it visits a state in $\mathcal{S}_- = \mathcal{S} \setminus \mathcal{S}_+$. That is, if the CTMC stays at state $i \in \mathcal{S}_+$ or the CTMC stays at state $i \in \mathcal{S}_-$ and $Y(t) > 0$, then the fluid level changes at rate r_i,

$$\frac{d}{dt}Y(t) = r_i.$$

That is, if $i \in \mathcal{S}_+$, then $r_i > 0$ and the fluid level increases, if $i \in \mathcal{S}_-$ and $Y(t) > 0$, then $r_i < 0$ and the fluid level decreases. If $i \in \mathcal{S}_-$ and $Y(t) = 0$, then the fluid level does not change.

The characterizing matrices of the MFQ are the generator matrix of the CTMC, $\mathbf{Q}$, and the diagonal matrix of the fluid rates, $\mathbf{R}$. The rate dependent decomposition of the characterizing matrices are

$$\mathbf{Q} = \begin{bmatrix} \mathbf{Q}_{++} & \mathbf{Q}_{+-} \\ \mathbf{Q}_{-+} & \mathbf{Q}_{--} \end{bmatrix} \quad \text{and} \quad \mathbf{R} = \begin{bmatrix} \mathbf{R}_+ & 0 \\ 0 & -\mathbf{R}_- \end{bmatrix}. \tag{1}$$

The stationary behaviour of the MFQ is characterized by the following measures: the empty buffer probability $\pi_i = \lim_{t\to\infty} Pr(J(t) = i, Y(t) = 0)$ and the fluid density $f_i(x) = \lim_{t\to\infty} \frac{d}{dx} Pr(J(t) = i, Y(t) < x)$. We note that $Pr(J(t) = i, Y(t) = 0) = 0$ for $i \in \mathcal{S}_+$, because the fluid buffer cannot be empty when the fluid rate is positive.

In this work, we do not look into the solution methods to compute these stationary measures, we only discuss the set of differential, boundary and normalizing equations which have to be satisfied by the stationary solution.

We also introduce the related quantities $\nu_i = \pi_i |r_i|$ and $\phi_i(x) = f_i(x)|r_i|$, which is often referred to as flux. The vectors composed of the state dependent measures are $\pi_- = [\pi_i]_{i\in\mathcal{S}_-}$, $\nu_- = [\nu_i]_{i\in\mathcal{S}_-}$, $f_-(x) = [f_i(x)]_{i\in\mathcal{S}_-}$, $f_+(x) = [f_i(x)]_{i\in\mathcal{S}_+}$, $\phi_-(x) = [\phi_i(x)]_{i\in\mathcal{S}_-}$, and $\phi_+(x) = [\phi_i(x)]_{i\in\mathcal{S}_+}$.

The stationary measures satisfy the following ordinary differential relations [17]

$$\frac{d}{dx} f_+(x)\mathbf{R}_+ = f_+(x)\mathbf{Q}_{++} + f_-(x)\mathbf{Q}_{-+}, \tag{2}$$

$$-\frac{d}{dx} f_-(x)\mathbf{R}_- = f_+(x)\mathbf{Q}_{+-} + f_-(x)\mathbf{Q}_{--}, \tag{3}$$

and, equivalently,

$$\frac{d}{dx} \phi_+(x) = \phi_+(x)\mathbf{T}_{++} + \phi_-(x)\mathbf{T}_{-+}, \tag{4}$$

$$-\frac{d}{dx} \phi_-(x) = \phi_+(x)\mathbf{T}_{+-} + \phi_-(x)\mathbf{T}_{--}, \tag{5}$$

where $\mathbf{T} = |\mathbf{R}^{-1}|\mathbf{Q}$. The initial conditions of these ordinary differential equations [17] are

$$f_+(0)\mathbf{R}_+ = \pi\mathbf{Q}_{-+}, \tag{6}$$

$$-f_-(0)\mathbf{R}_- = \pi\mathbf{Q}_{--}, \tag{7}$$

and, equivalently, we have

$$\phi_+(0) = \nu\mathbf{T}_{-+}, \tag{8}$$

$$-\phi_-(0) = \nu\mathbf{T}_{--}. \tag{9}$$

Finally, the normalizing equations for the density and the flux are

$$1 = \pi \mathbb{1} + \int_0^\infty \left(f_+(x)\mathbb{1} + f_-(x)\mathbb{1} \right) dx \quad \text{and} \tag{10}$$

$$1 = \nu \mathbf{R}_-^{-1}\mathbb{1} + \int_0^\infty \left(\phi_+(x)\mathbf{R}_+^{-1}\mathbb{1} + \phi_-(x)\mathbf{R}_-^{-1}\mathbb{1} \right) dx.$$

2.2 QBD Process

Discrete or continuous time QBD processes are discrete or continuous time Markov chains whose states can be efficiently represented by two discrete variables $\{\mathcal{X}, \mathcal{J}\}$, where $\mathcal{X} \in \{0, 1, \dots\}$ is called the level and $\mathcal{J} \in \{1, 2, \dots, N\}$ is called the phase. The direct state transitions of QBD processes are restricted between states of the same level or the neighbouring levels.

We assume (level) homogeneous QBD processes, where matrix $\mathbf{B}$ holds the rates of the level backward transitions, $\mathbf{F}$ the rates of the level forward transitions, and $\mathbf{L}$ the ones of the local transitions, which are not accompanied by the change of the level. At level zero the behavior of the local transitions can differ from the regular ones and the matrix describining these transitions is denoted by $\mathbf{L}'$.

In case of discrete time QBD processes, the one step state transition probability matrix has the following block tri-diagonal structure

$$\mathbf{P} = \begin{pmatrix} \mathbf{L}' & \mathbf{F} & & \\ \mathbf{B} & \mathbf{L} & \mathbf{F} & \\ & \mathbf{B} & \mathbf{L} & \mathbf{F} \\ & & \ddots & \ddots & \ddots \end{pmatrix}.$$

The stationary distribution of this QBD satisfies the stationary equations

$$p_0 = p_0 \mathbf{L}' + p_1 \mathbf{B}, \tag{11}$$

$$p_n = p_{n-1}\mathbf{F} + p_n \mathbf{L} + p_{n+1}\mathbf{B} \quad \text{for } n \geq 1, \tag{12}$$

where p_n is the stationary probability vector associated with level n.

3 Relation of MFQs and QBD Process with Singe Buffer

3.1 QBD Structure Proposed in [1]

Ahn and Ramaswami established a relationship between QBD processes of a given structure and MFQs. In their QBD process, the states of the background CTMC of the MFQ are mapped to the phases of the QBD. Following the $\mathcal{S}_+$, $\mathcal{S}_-$ partitioning of the states, the transition matrices of the QBD process proposed in [1] are as follows:

$$\mathbf{B} = \frac{1}{2} \begin{bmatrix} 0 & 0 \\ 0 & \mathbf{I} \end{bmatrix}, \mathbf{L} = \frac{1}{2} \begin{bmatrix} \mathbf{I} & 0 \\ \mathbf{P}_{-+} & \mathbf{P}_{--} \end{bmatrix}, \mathbf{F} = \frac{1}{2} \begin{bmatrix} \mathbf{P}_{++} & \mathbf{P}_{+-} \\ 0 & 0 \end{bmatrix}. \tag{13}$$

For level 0 the local matrix is $\mathbf{L}' = \mathbf{L}+\mathbf{B}$ and matrix $\mathbf{P}$ is defined as $\mathbf{P} = \mathbf{T}/\lambda+\mathbf{I}$, where $\mathbf{T} = |\mathbf{R}^{-1}|\mathbf{Q}$ and $\lambda = \max_{i,j} |T_{i,j}|$.

Based on the stochastic interpretation of the QBD process with these transition matrices and the MFQ characterized by $\mathbf{Q}$ and $\mathbf{R}$, the stationary solution of the MFQ is provided based on the stationary solution of the QBD process. Unfortunately, this solution requires a different scaling of time in $\mathcal{S}_+$ and $\mathcal{S}_-$ [1, Theorem 4]. It is also mentioned in [1] that different QBD structures can be used for establishing such relation between a QBD process and a MFQ. The modified structure we use in this paper allows identical scaling of time in $\mathcal{S}_+$ and $\mathcal{S}_-$.

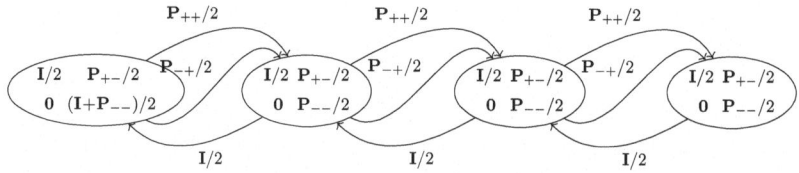

Fig. 1. Block structure of the QBD

3.2   QBD Process with Modified Structure

In this work, we study a slightly modified QBD structure, whose analytical treatment is simpler. Let $X(t) = \{N(t), J(t)\}$ be a QBD process where $J(t) \in \mathcal{S}$ is the state of the background CTMC and $N(t)$ is the level of the QBD at time t. Based on the $\mathcal{S}_+ - \mathcal{S}_-$ decomposition, the blocks structure of the characterizing matrices of the QBD process are

$$\mathbf{B} = \frac{1}{2}\begin{bmatrix} \mathbf{0} & \mathbf{0} \\ \mathbf{0} & \mathbf{I} \end{bmatrix}, \mathbf{L} = \frac{1}{2}\begin{bmatrix} \mathbf{I} & \mathbf{P}_{+-} \\ \mathbf{0} & \mathbf{P}_{--} \end{bmatrix}, \mathbf{F} = \frac{1}{2}\begin{bmatrix} \mathbf{P}_{++} & \mathbf{0} \\ \mathbf{P}_{-+} & \mathbf{0} \end{bmatrix}. \tag{14}$$

For level 0 the generator matrix is $\mathbf{L}' = \mathbf{L} + \mathbf{B}$. The state transition structure of the QBD is depicted in Fig. 1. Arrows from/to the upper half of the ellipses indicate transitions from/to $\mathcal{S}^+$.

Let us define $p_{n,i} = \lim_{t\to\infty} Pr(N(t) = n, J(t) = i)$, which describes the stationary distribution of the QBD. Let p_n^+ and p_n^- be the row vectors composed of $p_{n,i}$ for $i \in \mathcal{S}^+$ and $i \in \mathcal{S}^-$, respectively. These vectors satisfy the following stationary equations for $n \geq 1$

$$p_0^+ = 0, \tag{15}$$

$$p_0^- = p_0^-(\mathbf{I}+\mathbf{P}_{--})/2 + p_1^-\mathbf{I}/2, \tag{16}$$

$$p_n^+ = p_{n-1}^+\mathbf{P}_{++}/2 + p_{n-1}^-\mathbf{P}_{-+}/2 + p_n^+\mathbf{I}/2, \tag{17}$$

$$p_n^- = p_n^-\mathbf{P}_{--}/2 + p_n^+\mathbf{P}_{+-}/2 + p_{n+1}^-\mathbf{I}/2, \tag{18}$$

which can be simplified to

$$p_1^- = p_0^- (\mathbf{I} - \mathbf{P}_{--}), \tag{19}$$

$$p_n^+ = p_{n-1}^+ \mathbf{P}_{++} + p_{n-1}^- \mathbf{P}_{-+}, \tag{20}$$

$$p_n^- (2\mathbf{I} - \mathbf{P}_{--}) = p_n^+ \mathbf{P}_{+-} + p_{n+1}^- \mathbf{I}. \tag{21}$$

3.3   Algebraic Proof of the Relation of MFQs and QBD Processes

The following theorem relates the stationary behaviour of the QBD processes defined in (14) with $\mathbf{L}' = \mathbf{L} + \mathbf{B}$ and the MFQ with characterizing matrices defined in (1).

Theorem 1. *When* $\mathbf{T} = |\mathbf{R}^{-1}|\mathbf{Q}$, $\lambda = \max_{i,j} |T_{i,j}|$, $\mathbf{P} = \mathbf{T}/\lambda + \mathbf{I}$, *and* p_n^+, p_n^- *is a non-zero solution of* (19)–(21), *then*

$$\hat{\phi}_\pm(x) = \sum_{n=1}^{\infty} \frac{\lambda^n x^{n-1} e^{-\lambda x}}{(n-1)!} p_n^\pm \quad and \quad \hat{\nu} = p_0^- \tag{22}$$

satisfy the differential Eqs. (4) *and* (5) *with boundary conditions* (8) *and* (9).

The theorem states that the solution of the MFQ is a mixture of Erlang distributions of order n and rate λ weighted according to the stationary distribution of level n of the QBD process for $n \geq 1$ and the empty buffer probability of the MFQ is related to the stationary distribution of level 0 of the QBD process.

Proof. When $\mathbf{P} = \mathbf{T}/\lambda + \mathbf{I}$, we have $\mathbf{P}_{++} = \mathbf{T}_{++}/\lambda + \mathbf{I}$, $\mathbf{P}_{+-} = \mathbf{T}_{+-}/\lambda$, $\mathbf{P}_{--} = \mathbf{T}_{--}/\lambda + \mathbf{I}$, $\mathbf{P}_{-+} = \mathbf{T}_{-+}/\lambda$.

Substituting this into (19)–(21), for $n \geq 1$, we get

$$\lambda p_1^- = -p_0^- \mathbf{T}_{--}, \tag{23}$$

$$\lambda p_i^+ = p_{i-1}^+ (\mathbf{T}_{++} + \lambda \mathbf{I}) + p_{i-1}^- \mathbf{T}_{-+}, \tag{24}$$

$$p_i^- (\lambda \mathbf{I} - \mathbf{T}_{--}) = p_i^+ \mathbf{T}_{+-} + p_{i+1}^- \lambda \mathbf{I}. \tag{25}$$

For $i \geq 1$, we have

$$\frac{d}{dx} \frac{\lambda^i x^{i-1} e^{-\lambda x}}{(i-1)!} = \mathcal{I}_{\{i>1\}} \frac{\lambda^i x^{i-2} e^{-\lambda x}}{(i-2)!} - \frac{\lambda^{i+1} x^{i-1} e^{-\lambda x}}{(i-1)!}, \tag{26}$$

from which $\hat{\phi}_\pm(x)$ satisfies

$$\frac{d}{dx} \hat{\phi}_\pm(x) = \sum_{i=2}^{\infty} \frac{\lambda^i x^{i-2} e^{-\lambda x}}{(i-2)!} p_i^\pm - \lambda \sum_{i=1}^{\infty} \frac{\lambda^i x^{i-1} e^{-\lambda x}}{(i-1)!} p_i^\pm \tag{27}$$

$$= \lambda \sum_{i=1}^{\infty} \frac{\lambda^i x^{i-1} e^{-\lambda x}}{(i-1)!} p_{i+1}^\pm - \lambda \hat{\phi}_\pm(x).$$

Multiplying (24) by $\frac{\lambda^{i-1}x^{i-2}e^{-\lambda x}}{(i-2)!}$ and summing up from $i = 2$ to ∞ gives

$$\lambda \sum_{i=2}^{\infty} \frac{\lambda^{i-1}x^{i-2}e^{-\lambda x}}{(i-2)!} p_i^+ \tag{28}$$

$$= \sum_{i=2}^{\infty} \frac{\lambda^{i-1}x^{i-2}e^{-\lambda x}}{(i-2)!} \left(p_{i-1}^+ (\mathbf{T}_{++} + \lambda \mathbf{I}) + p_{i-1}^- \mathbf{T}_{-+} \right), \text{ and}$$

$$\frac{d}{dx}\hat{\phi}_+(x) + \lambda\hat{\phi}_+(x) = \hat{\phi}_+(x)(\mathbf{T}_{++} + \lambda\mathbf{I}) + \hat{\phi}_-(x)\mathbf{T}_{-+}, \text{ and}$$

$$\frac{d}{dx}\hat{\phi}_+(x) = \hat{\phi}_+(x)\mathbf{T}_{++} + \hat{\phi}_-(x)\mathbf{T}_{-+}.$$

Multiplying (25) by $\frac{\lambda^{i}x^{i-1}e^{-\lambda x}}{(i-1)!}$ and summing up from $i = 1$ to ∞ gives

$$\sum_{i=1}^{\infty} \frac{\lambda^{i}x^{i-1}e^{-\lambda x}}{(i-1)!} p_i^- (\lambda\mathbf{I} - \mathbf{T}_{--}) = \sum_{i=1}^{\infty} \frac{\lambda^{i}x^{i-1}e^{-\lambda x}}{(i-1)!} \left(p_i^+ \mathbf{T}_{+-} + p_{i+1}^- \lambda\mathbf{I} \right), \text{ and}$$

$$\tag{29}$$

$$\hat{\phi}_-(x)(\lambda\mathbf{I} - \mathbf{T}_{--}) = \hat{\phi}_+(x)\mathbf{T}_{+-} + \frac{d}{dx}\hat{\phi}_-(x) + \lambda\hat{\phi}_-(x), \text{ and}$$

$$-\frac{d}{dx}\hat{\phi}_-(x) = \hat{\phi}_+(x)\mathbf{T}_{+-} + \hat{\phi}_-(x)\mathbf{T}_{--},$$

where, from (28), we used that

$$\sum_{i=2}^{\infty} \frac{\lambda^{i}x^{i-2}e^{-\lambda x}}{(i-2)!} p_i^\pm = \lambda \sum_{i=1}^{\infty} \frac{\lambda^{i}x^{i-1}e^{-\lambda x}}{(i-1)!} p_{i+1}^\pm = \frac{d}{dx}\hat{\phi}_\pm(x) + \lambda\hat{\phi}_\pm(x). \tag{30}$$

For the initial conditions, we start from the definition of $\hat{\phi}_\pm(x)$ given in (22), from which

$$\hat{\phi}_\pm(0) = \lambda p_1^\pm. \tag{31}$$

Substituting p_1^- from (23) and p_1^+ from (24) (and using that $\hat{\nu} = p_0^-$ and $p_0^+ = 0$) gives

$$\hat{\phi}_+(0) = \hat{\nu}\mathbf{T}_{-+}, \tag{32}$$

$$-\hat{\phi}_-(0) = \hat{\nu}\mathbf{T}_{--}. \tag{33}$$

□

Theorem 1 does not imply that $\phi_\pm(x) = \hat{\phi}_\pm(x)$ and $\nu = \hat{\nu}$, because the normalizing condition of the MFQ and the QBD process differ. $\phi(x)$ and ν satisfy the normalizing Eq. (10), while $\hat{\phi}(x)$ and $\hat{\nu}$ are normalized as follows

$$\sum_{i=0}^{\infty}(p_i^+\mathbb{1} + p_i^-\mathbb{1}) = \hat{\nu} + \int_{x=0}^{\infty} \hat{\phi}_+(x)\mathbb{1} + \hat{\phi}_-(x)\mathbb{1}dx = 1. \tag{34}$$

4   Two Fluid Buffers

In this section we investigate if the relation of QBD processes and MFQs can be extended for simple MFQs with two buffers using the same approach as in the previous section.

4.1   Markov Fluid Queue

We consider a MFQ with two infinite buffers $(J(t), Y_1(t), Y_2(t))$, where $J(t)$ is the state of the background CTMC and $Y_i(t)$ is the fluid level of buffer i ($i \in \{1, 2\}$) at time t. The state space of the CTMC is composed of two disjoint subsets $\mathcal{S}_+$ and $\mathcal{S}_- = \mathcal{S} \setminus \mathcal{S}_+$ such that the fluid level of both buffers increases at rate 1 in $\mathcal{S}_+$ and in $\mathcal{S}_-$, the fluid level of buffer 1 decreases with rate 1 and the fluid level of buffer 2 decreases with rate $r_2 < 1$, if the buffers are non-empty. That is, the characterizing matrices of the MFQ are $\mathbf{Q} = \begin{bmatrix} \mathbf{Q}_{++} & \mathbf{Q}_{+-} \\ \mathbf{Q}_{-+} & \mathbf{Q}_{--} \end{bmatrix}$, $\mathbf{R}_1 = \begin{bmatrix} \mathbf{I} & 0 \\ 0 & -\mathbf{I} \end{bmatrix}$ and $\mathbf{R}_2 = \begin{bmatrix} \mathbf{I} & 0 \\ 0 & -\mathbf{r_2 I} \end{bmatrix}$. A trajectory of the fluid levels is depicted in Fig. 2. An important consequence of this model behaviour is that $Y_1(t) \leq Y_2(t)$ for $t > 0$. As a consequence, the MFQ is stable if buffer 2 is stable. Figure 2 indicates 4 possible cases:

- $Y_2(t) > Y_1(t) > 0$,
- $Y_2(t) > 0, Y_1(t) = 0$,
- $Y_2(t) = Y_1(t) = 0$,
- $Y_2(t) = Y_1(t) > 0$.

The stationary measures associated with these 4 cases are as follows

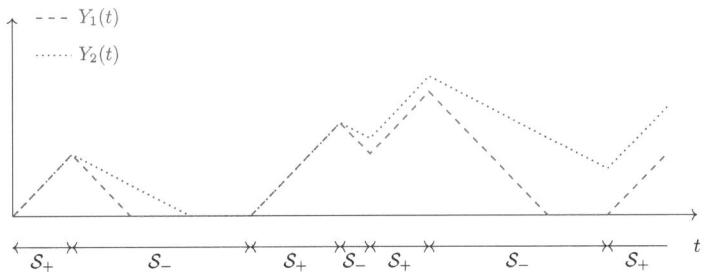

Fig. 2. Evolution of the buffer contents with $r_2 = 0.5$

$$W_i(x, y) = \lim_{t \to \infty} \frac{d}{dx} \frac{d}{dy} Pr(J(t) = i, Y_1(t) < x, Y_2(t) < y),$$

$$U_i(y) = \lim_{t \to \infty} \frac{d}{dy} Pr(J(t) = i, Y_1(t) = 0, Y_2(t) < y),$$

$$\pi_i = \lim_{t \to \infty} Pr(J(t) = i, Y_1(t) = Y_2(t) = 0),$$

$$V_i(x) = \lim_{t \to \infty} \frac{d}{dx} Pr(J(t) = i, Y_1(t) = Y_2(t) < x).$$

The vectors composed of the state dependent measures are $\pi = [\pi_i]_{i \in \mathcal{S}_-}$, $U(y) = [U_i(y)]_{i \in \mathcal{S}_-}$, $V(x) = [V_i(x)]_{i \in \mathcal{S}_+}$, $W_+(x, y) = [W_i(x, y)]_{i \in \mathcal{S}_+}$, and $W_-(x, y) = [W_i(x, y)]_{i \in \mathcal{S}_-}$.

The stationary solution satisfies the following equations

$$\pi^+ = \mathbf{0} ; \qquad \mathbf{0} = \pi^- \mathbf{Q}_{--} + r_2 U(0); \tag{35}$$

$$\frac{\partial}{\partial x} V(x) = V(x) \mathbf{Q}_{++}, \tag{36}$$

with initial condition

$$V(0) = \pi^- \mathbf{Q}_{-+}; \tag{37}$$

$$-r_2 \frac{\partial}{\partial y} U(y) = U(y) \mathbf{Q}_{--} - W^-(0, y); \tag{38}$$

$$\frac{\partial}{\partial x} W^+(x, y) + \frac{\partial}{\partial y} W^+(x, y) = W^+(x, y) \mathbf{Q}_{++} + W^-(x, y) \mathbf{Q}_{-+}, \tag{39}$$

with initial conditions

$$W^+(x, x) = \mathbf{0} \quad \text{and} \quad W^+(0, y) = U(y) \mathbf{Q}_{-+}; \tag{40}$$

$$-\frac{\partial}{\partial x} W^-(x, y) - r_2 \frac{\partial}{\partial y} W^-(x, y) = W^+(x, y) \mathbf{Q}_{+-} + W^-(x, y) \mathbf{Q}_{--}, \tag{41}$$

with initial condition

$$r_2 W^-(x, x) = V^+(x) \mathbf{Q}_{+-}. \tag{42}$$

4.2   Level Dependent Quasi Birth Death Process

In this section we introduce a level dependent quasi birth death (LDQBD) process whose structure is meant to represent the behaviour of the MFQ with two buffers and we check if the stationary behaviour of the MFQ with two buffers and the LDQBD process are related.

The block structure of the process is depicted in Fig. 3 applying the same graphical representations of the transitions associated with $\mathcal{S}^+$ and $\mathcal{S}^-$ as in Fig. 1, that is, the arrows from/to the upper half of the ellipses indicate transitions from/to $\mathcal{S}^+$ and the lower half of the ellipses are related to $\mathcal{S}^-$.

Assuming, that the blocks of states of this process (indicated by ellipses in the figure) are such that the associated *levels* increase along the horizontal axis, and all of the blocks along a vertical line compose the *phases* of the given level, the obtained stochastic process is a LDQBD process whose state transitions can be described with the transition probability matrix

$$\mathbf{L} = \begin{pmatrix} \mathbf{L}^{(0)} & \mathbf{F}^{(0)} & & & \\ \mathbf{B}^{(1)} & \mathbf{L}^{(1)} & \mathbf{F}^{(1)} & & \\ & \mathbf{B}^{(2)} & \mathbf{L}^{(2)} & \mathbf{F}^{(2)} & \\ & & \ddots & \ddots & \ddots \end{pmatrix}, \tag{43}$$

where the size of the matrices of the different levels increases level-by-level. For the detailed internal structure of the non-zero blocks of (43) we refer to Fig. 3.

We decompose the stationary probability vector of the LDQBD into the following blocks: $p_{00}^+, p_{00}^-, p_{10}^+, p_{10}^-, p_{11}^+, p_{11}^-, p_{12}^+, p_{12}^-, \ldots, p_{i0}^+, p_{i0}^-, \ldots, p_{i,i+1}^+, p_{i,i+1}^-, p_{i+1,0}^+, p_{i+1,0}^-, \ldots$. The stationary probability of the transient states of the LDQBD is zero, from which

$$p_{i-1,0}^+ = 0 \text{ and } p_{i,i+1}^- = 0 \quad \text{for } i \geq 1. \tag{44}$$

According to Fig. 3, for $i \geq 1$ and $i \geq j \geq 1$, the decomposed vectors satisfy the following stationary equations

$$p_{00}^- = p_{00}^-(\mathbf{I}+\mathbf{P}_{--})/2 + (p_{10}^- + p_{11}^-)r_2\mathbf{I}/2, \tag{45}$$

$$p_{i0}^- = p_{i0}^-((1-r_2)\mathbf{I}+\mathbf{P}_{--})/2 + p_{i1}^-(1-r_2)\mathbf{I}/2 + (p_{i+1,0}^- + p_{i+1,1}^-)r_2\mathbf{I}/2, \tag{46}$$

$$p_{ij}^+ = p_{ij}^+\mathbf{I}/2 + p_{i-1,j-1}^+\mathbf{P}_{++}/2 + p_{i-1,j-1}^-\mathbf{P}_{-+}/2, \tag{47}$$

$$p_{ij}^- = p_{ij}^-\mathbf{P}_{--}/2 + p_{ij}^+\mathbf{P}_{+-}/2 + p_{i+1,j+1}^-r_2\mathbf{I}/2 \tag{48}$$

$$+ \mathcal{I}_{\{j<i\}}p_{i,j+1}^-(1-r_2)\mathbf{I}/2 + \mathcal{I}_{\{j=i\}}p_{i,j+1}^-\mathbf{P}_{+-}/2,$$

$$p_{1,2}^+ = p_{1,2}^+\mathbf{I}/2 + p_{0,0}^-\mathbf{P}_{-+}/2, \tag{49}$$

$$p_{i+1,i+2}^+ = p_{i+1,i+2}^+\mathbf{I}/2 + p_{i,i+1}^+\mathbf{P}_{++}/2. \tag{50}$$

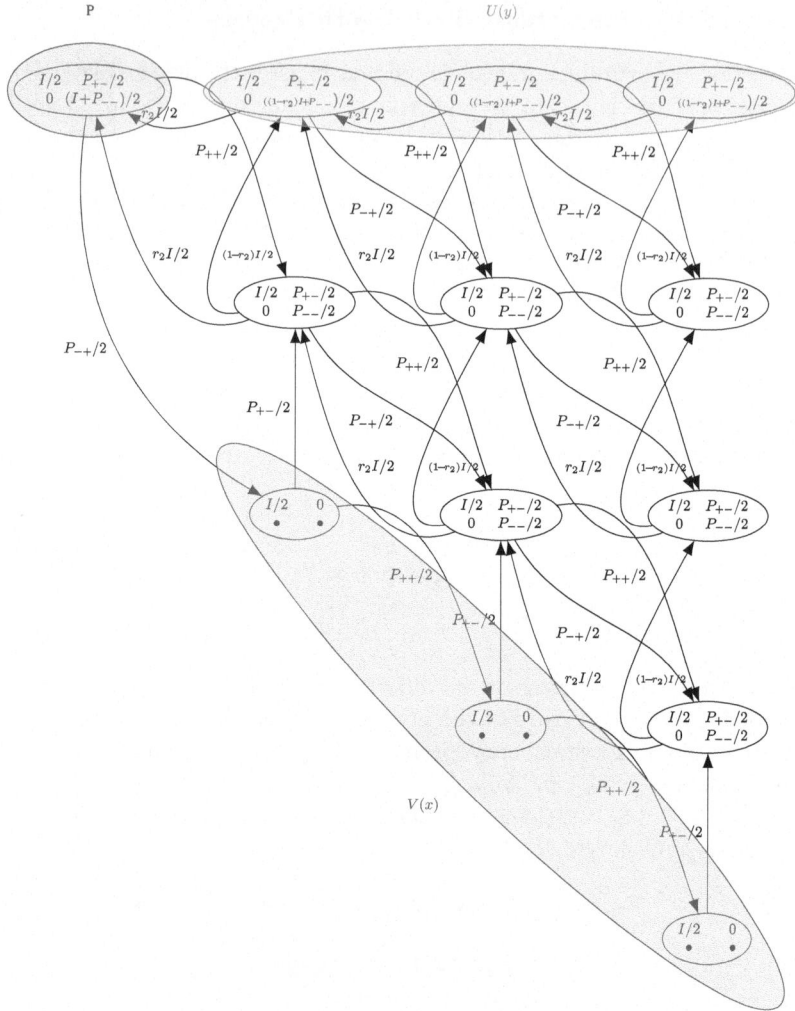

Fig. 3. Block structure of the LDQBD

4.3   Relation of the MFQ and the LDQBD

In this section we follow the same approach as in Theorem 1 and check if the stationary behaviour of the LDQBD is related to the one of the MFQ.

Theorem 2. *Assuming* $\mathbf{P} = \mathbf{Q}/\lambda + \mathbf{I}$, $\tilde{\pi} = p_{0,0}^-$, $\tilde{V}(x) = \sum_{i=1}^{\infty} \frac{\lambda^i x^{i-1} e^{-\lambda x}}{(i-1)!} p_{i,i+1}^+$, $\tilde{U}(y) = \sum_{i=1}^{\infty} \frac{\lambda^i y^{i-1} e^{-\lambda y}}{(i-1)!} p_{i,0}^-$, $\tilde{W}(x,y)^{\pm} = \sum_{i=1}^{\infty} \sum_{j=1}^{i} \frac{\lambda^i y^{i-1} e^{-\lambda y}}{(i-1)!} \frac{\lambda^j x^{j-1} e^{-\lambda x}}{(j-1)!} p_{i,j}^{\pm}$, $\tilde{V}(x)$, $\tilde{U}(x)$ and $\tilde{W}(x,y)$ satisfy*

$$\frac{d}{dx}\tilde{V}(x) = \tilde{V}(x)\mathbf{Q}_{++}. \tag{51}$$

$$-r_2 \frac{d}{dy} \tilde{U}(y) = \tilde{U}(y) \mathbf{Q}_{--} + \underbrace{(1-r_2)\tilde{W}(0,y)^- + r_2 \left(\frac{d}{dx}\tilde{W}(0,y)^- + \lambda \tilde{W}(0,y)^- \right)}_{different\ from\ MFQ},$$

$$\tag{52}$$

$$\frac{\partial}{\partial y}\tilde{W}(x,y)^+ + \frac{\partial}{\partial x}\tilde{W}(x,y)^+ = \tilde{W}(x,y)^+ \mathbf{Q}_{++} + \tilde{W}(x,y)^- \mathbf{Q}_{-+}$$
$$\underbrace{- \frac{1}{\lambda} \frac{\partial}{\partial x} \frac{\partial}{\partial y}\tilde{W}(x,y)^+}_{different\ from\ MFQ}, \tag{53}$$

$$-\frac{\partial}{\partial x}\tilde{W}(x,y)^- - r_2 \frac{\partial}{\partial y}\tilde{W}(x,y)^- = \tilde{W}(x,y)^- \mathbf{Q}_{--} + \tilde{W}(x,y)^- \mathbf{Q}_{--}$$
$$\underbrace{+ \varepsilon(x,y)\mathbf{Q}_{+-} + \frac{r_2}{\lambda} \frac{\partial}{\partial x} \frac{\partial}{\partial y}\tilde{W}(x,y)^-}_{different\ from\ MFQ}, \quad \tag{54}$$

where $\varepsilon(x,y) = \sum_{i=2}^{\infty} \frac{\lambda^{i-1}x^{i-2}e^{-\lambda x}}{(i-2)!} \frac{\lambda^{i-1}y^{i-2}e^{-\lambda y}}{(i-2)!} p_{i-1,i}^-.$

The proof is provided in the appendix.

While $\tilde{V}(x)$ satisfies the same differential equation as $V(x)$ in (36), unfortunately, $\tilde{U}(x)$, $\tilde{W}(x,y)^+$ and $\tilde{W}(x,y)^-$ are characterized by different differential equations than (38), (39) and (41), respectively. It means that the stationary distribution of the MFQ with two buffers cannot be established based on this LDQBD.

5   Conclusion

We revisited the relation of MFQs and QBD processes in order to extend it for MFQs with two buffers. To this end, we replaced the stochastic intuition based discussion of [1] with an algebraic one and provided an algebraic proof of the relation of MFQs and QBD processes. For the analysis of a simple MFQ with two buffers we introduced a LDQBD, whose structure mimics the behaviour of the queue in the same way as in the single buffer case. Unfortunately, the accurate algebraic approach indicated that the stationary behaviour of the MFQs with two buffers and the LDQBD process differ.

A   Proof of Theorem 2

Using $\mathbf{P}_{++} = \mathbf{Q}_{++}/\lambda + \mathbf{I}$, $\mathbf{P}_{+-} = \mathbf{Q}_{+-}/\lambda$, $\mathbf{P}_{--} = \mathbf{Q}_{--}/\lambda + \mathbf{I}$, $\mathbf{P}_{-+} = \mathbf{Q}_{-+}/\lambda$, for $i \geq j \geq 1$, Equations (45)–(50) can be simplified to

$$-p_{00}^- \mathbf{Q}_{--} = \lambda r_2 (p_{10}^- + p_{11}^-), \tag{55}$$

$$p_{i0}^-(\lambda r_2 \mathbf{I} - \mathbf{Q}_{--}) = \lambda(1-r_2)p_{i1}^- + \lambda r_2 (p_{i+1,0}^- + p_{i+1,1}^-), \tag{56}$$

$$\lambda p_{ij}^+ = p_{i-1,j-1}^+(\lambda\mathbf{I} + \mathbf{Q}_{++}) + p_{i-1,j-1}^-\mathbf{Q}_{-+}, \tag{57}$$

$$p_{ij}^-(\lambda\mathbf{I}-\mathbf{Q}_{--}) = p_{ij}^+\mathbf{Q}_{+-} + \lambda r_2 p_{i+1,j+1}^-$$
$$+ \mathcal{I}_{\{j<i\}}\lambda(1-r_2)p_{i,j+1}^- + \mathcal{I}_{\{j=i\}}p_{i,j+1}^-\mathbf{Q}_{+-}, \tag{58}$$

$$\lambda p_{1,2}^+ = p_{0,0}^-\mathbf{Q}_{-+}, \tag{59}$$

$$\lambda p_{i+1,i+2}^+ = p_{i,i+1}^+(\lambda\mathbf{I} + \mathbf{Q}_{++}). \tag{60}$$

Multiplying (60) by $\frac{\lambda^i x^{i-1} e^{-\lambda x}}{(i-1)!}$ and summing up from $i = 1$ to ∞ gives

$$\sum_{i=1}^{\infty} \frac{\lambda^i x^{i-1} e^{-\lambda x}}{(i-1)!}\lambda p_{i+1,i+2}^+ = \sum_{i=1}^{\infty} \frac{\lambda^i x^{i-1} e^{-\lambda x}}{(i-1)!} p_{i,i+1}^+(\lambda\mathbf{I} + \mathbf{Q}_{++}), \tag{61}$$

$$\frac{d}{dx}\tilde{V}(x) + \lambda\tilde{V}(x) = \tilde{V}(x)(\lambda\mathbf{I} + \mathbf{Q}_{++}), \tag{62}$$

which results in (51).

By definition $\tilde{W}(0,x)^{\pm} = \sum_{i=1}^{\infty} \frac{\lambda^i x^{i-1} e^{-\lambda x}}{(i-1)!}\lambda p_{i,1}^{\pm}$. Multiplying (56) by $\frac{\lambda^i x^{i-1} e^{-\lambda x}}{(i-1)!}$ and summing up from $i = 1$ to ∞ gives

$$\sum_{i=1}^{\infty} \frac{\lambda^i x^{i-1} e^{-\lambda x}}{(i-1)!} p_{i0}^-(\lambda r_2\mathbf{I}-\mathbf{Q}_{--}) = \sum_{i=1}^{\infty} \frac{\lambda^i x^{i-1} e^{-\lambda x}}{(i-1)!}\left(\lambda(1-r_2)p_{i1}^- + \lambda r_2(p_{i+1,0}^- + p_{i+1,1}^-)\right),$$

$$\tilde{U}(x)(\lambda r_2\mathbf{I}-\mathbf{Q}_{--}) = \underbrace{\sum_{i=1}^{\infty} \frac{\lambda^i x^{i-1} e^{-\lambda x}}{(i-1)!}\lambda(1-r_2)p_{i1}^-}_{(1-r_2)\tilde{W}(0,x)^-} + \underbrace{\sum_{i=1}^{\infty} \frac{\lambda^i x^{i-1} e^{-\lambda x}}{(i-1)!}\lambda r_2 p_{i+1,1}^-}_{r_2\left(\frac{d}{dx}\tilde{W}(0,x)^- + \lambda\tilde{W}(0,x)^-\right)}$$

$$+ \underbrace{\sum_{i=1}^{\infty} \frac{\lambda^i x^{i-1} e^{-\lambda x}}{(i-1)!}\lambda r_2 p_{i+1,0}^-}_{r_2\left(\frac{d}{dx}\tilde{U}(x) + \lambda\tilde{U}(x)\right)},$$

which results in (51).

For the computation of $\tilde{W}(x,y)$ we need the following lemma.

Lemma 1. *The derivatives of* $\tilde{W}(x,y)^{\pm} = \sum_{i=1}^{\infty}\sum_{j=1}^{i} \frac{\lambda^i y^{i-1} e^{-\lambda y}}{(i-1)!} \frac{\lambda^j x^{j-1} e^{-\lambda x}}{(j-1)!} p_{i,j}^{\pm}$ *satisfy*

$$\frac{\partial}{\partial y}\tilde{W}(x,y)^{\pm} + \lambda\tilde{W}(x,y)^{\pm} = \sum_{i=2}^{\infty}\sum_{j=1}^{i} \frac{\lambda^i y^{i-2} e^{-\lambda y}}{(i-2)!} \frac{\lambda^j x^{j-1} e^{-\lambda x}}{(j-1)!} p_{i,j}^{\pm}, \tag{63}$$

$$\frac{\partial}{\partial x}\tilde{W}(x,y)^{\pm} + \lambda\tilde{W}(x,y)^{\pm} = \sum_{i=1}^{\infty}\sum_{j=2}^{i}\frac{\lambda^i y^{i-1}e^{-\lambda y}}{(i-1)!}\frac{\lambda^j x^{j-2}e^{-\lambda x}}{(j-2)!}p_{i,j}^{\pm} \tag{64}$$

$$= \sum_{i=2}^{\infty}\sum_{j=2}^{i-1}\frac{\lambda^{i-1} y^{i-2}e^{-\lambda y}}{(i-2)!}\frac{\lambda^j x^{j-2}e^{-\lambda x}}{(j-2)!}p_{i-1,j}^{\pm}, \tag{65}$$

and

$$\frac{\partial}{\partial x}\frac{\partial}{\partial y}\tilde{W}(x,y)^{\pm} + \lambda\frac{\partial}{\partial y}\tilde{W}(x,y)^{\pm} + \lambda\frac{\partial}{\partial x}\tilde{W}(x,y)^{\pm} + \lambda^2\tilde{W}(x,y)^{\pm} \tag{66}$$

$$= \sum_{i=2}^{\infty}\sum_{j=2}^{i}\frac{\lambda^i y^{i-2}e^{-\lambda y}}{(i-2)!}\frac{\lambda^j x^{j-2}e^{-\lambda x}}{(j-2)!}p_{i,j}^{\pm}.$$

Proof. The statements of the lemma can be obtained by substituting the definition of $\tilde{W}(x,y)^{\pm}$. We omit the details of the proof here.

Multiplying (57) by $\frac{\lambda^{i-1}y^{i-2}e^{-\lambda y}}{(i-2)!}\frac{\lambda^{j-1}x^{j-2}e^{-\lambda x}}{(j-2)!}$ and summing up from $i = 2$ to ∞ and $j = 2$ to i and utilizing (66) gives

$$\underbrace{\sum_{i=2}^{\infty}\sum_{j=2}^{i}\frac{\lambda^{i-1} y^{i-2}e^{-\lambda y}}{(i-2)!}\frac{\lambda^{j-1} x^{j-2}e^{-\lambda x}}{(j-2)!}\lambda p_{ij}^{+}}_{\frac{1}{\lambda}\frac{\partial}{\partial x}\frac{\partial}{\partial y}\tilde{W}(x,y)^{+} + \frac{\partial}{\partial y}\tilde{W}(x,y)^{+} + \frac{\partial}{\partial x}\tilde{W}(x,y)^{+} + \lambda\tilde{W}(x,y)^{+}}$$

$$= \underbrace{\sum_{i=2}^{\infty}\sum_{j=2}^{i}\frac{\lambda^{i-1} y^{i-2}e^{-\lambda y}}{(i-2)!}\frac{\lambda^{j-1} x^{j-2}e^{-\lambda x}}{(j-2)!}\left(p_{i-1,j-1}^{+}(\lambda\mathbf{I} + \mathbf{Q}_{++}) + p_{i-1,j-1}^{-}\mathbf{Q}_{-+}\right)}_{\tilde{W}(x,y)^{+}(\lambda\mathbf{I} + \mathbf{Q}_{++}) + \tilde{W}(x,y)^{-}\mathbf{Q}_{-+}}$$

which results in (53).

For $2 \le j \le i$ we rewrite (58) as

$$p_{i-1,j-1}^{-}(\lambda\mathbf{I} - \mathbf{Q}_{--}) - p_{i-1,j-1}^{+}\mathbf{Q}_{+-} \tag{67}$$

$$= \lambda r_2 p_{i,j}^{-} + \mathcal{I}_{\{j<i\}}\lambda(1-r_2)p_{i-1,j}^{-} + \mathcal{I}_{\{j=i\}}p_{i-1,j}^{-}\mathbf{Q}_{+-},$$

Multiplying (67) by $\frac{\lambda^{i-1}y^{i-2}e^{-\lambda y}}{(i-2)!}\frac{\lambda^{j-1}x^{j-2}e^{-\lambda x}}{(j-2)!}$ and summing up from $i=2$ to ∞ and $j=2$ to i and utilizing Lemma 1 gives

$$\underbrace{\sum_{i=2}^{\infty}\sum_{j=2}^{i}\frac{\lambda^{i-1}y^{i-2}e^{-\lambda y}}{(i-2)!}\frac{\lambda^{j-1}x^{j-2}e^{-\lambda x}}{(j-2)!}p_{i-1,j-1}^{-}(\lambda\mathbf{I}-\mathbf{Q}_{--})}_{\tilde{W}(x,y)^{-}}$$

$$-\underbrace{\sum_{i=2}^{\infty}\sum_{j=2}^{i}\frac{\lambda^{i-1}y^{i-2}e^{-\lambda y}}{(i-2)!}\frac{\lambda^{j-1}x^{j-2}e^{-\lambda x}}{(j-2)!}p_{i-1,j-1}^{+}\mathbf{Q}_{+-}}_{\tilde{W}(x,y)^{+}}$$

$$=\underbrace{\sum_{i=2}^{\infty}\sum_{j=2}^{i}\frac{\lambda^{i-1}y^{i-2}e^{-\lambda y}}{(i-2)!}\frac{\lambda^{j-1}x^{j-2}e^{-\lambda x}}{(j-2)!}\lambda r_2 p_{i,j}^{-}}_{r_2\left(\frac{1}{\lambda}\frac{\partial}{\partial x}\frac{\partial}{\partial y}\tilde{W}(x,y)^{-}+\frac{\partial}{\partial y}\tilde{W}(x,y)^{-}+\frac{\partial}{\partial x}\tilde{W}(x,y)^{-}+\lambda\tilde{W}(x,y)^{-}\right)}$$

$$+\underbrace{\sum_{i=2}^{\infty}\sum_{j=2}^{i-1}\frac{\lambda^{i-1}y^{i-2}e^{-\lambda y}}{(i-2)!}\frac{\lambda^{j-1}x^{j-2}e^{-\lambda x}}{(j-2)!}\lambda(1-r_2)p_{i-1,j}^{-}}_{(1-r_2)\left(\frac{\partial}{\partial x}\tilde{W}(x,y)^{-}+\lambda\tilde{W}(x,y)^{-}\right)}$$

$$+\underbrace{\sum_{i=2}^{\infty}\frac{\lambda^{i-1}y^{i-2}e^{-\lambda y}}{(i-2)!}\frac{\lambda^{i-1}x^{i-2}e^{-\lambda x}}{(i-2)!}p_{i-1,i}^{-}\mathbf{Q}_{+-}}_{\varepsilon(x,y)},$$

that is

$$\tilde{W}(x,y)^{-}(\lambda\mathbf{I}-\mathbf{Q}_{--})-\tilde{W}(x,y)^{+}\mathbf{Q}_{+-}$$

$$=r_2\left(\frac{1}{\lambda}\frac{\partial}{\partial x}\frac{\partial}{\partial y}\tilde{W}(x,y)^{-}+\frac{\partial}{\partial y}\tilde{W}(x,y)^{-}+\frac{\partial}{\partial x}\tilde{W}(x,y)^{-}+\lambda\tilde{W}(x,y)^{-}\right)$$

$$+(1-r_2)\left(\frac{\partial}{\partial x}\tilde{W}(x,y)^{-}+\lambda\tilde{W}(x,y)^{-}\right)+\varepsilon(x,y)\mathbf{Q}_{+-},$$

which results in (54).

References

1. Ahn, S., Ramaswami, V.: Fluid flow models and queues - a connection by stochastic coupling. Stoch. Model. **19**(3), 325–348 (2003)
2. Akar, N., Sohraby, K.: An invariant subspace approach in $M/G/1$ and $G/M/1$ type Markov chains. Stoch. Model. **13**(3), 381–416 (1997)
3. Akar, N., Sohraby, K.: Infinite-and finite-buffer Markov fluid queues: a unified analysis. J. Appl. Probability, 557–569 (2004)
4. Anick, D., Mitra, D., Sondhi, M.M.: Stochastic theory of a data-handling system. Bell Sys. Thech. J. **61**(8), 1871–1894 (1982)
5. Asmussen, S.: Stationary distributions for fluid flow models with or without Brownian noise. Stoch. Model. **11**(1), 21–49 (1995)

6. Bean, N.G., Lewis, A., Nguyen, G.T., O'Reilly, M.M., Sunkara, V.: A discontinuous galerkin method for approximating the stationary distribution of stochastic fluid-fluid processes. Methodol. Comput. Appl. Probab. **24**(4), 2823–2864 (2022)
7. Bean, N.G., O'Reilly, M.M.: The stochastic fluid-fluid model: a stochastic fluid model driven by an uncountable-state process, which is a stochastic fluid model itself. Stochastic Processes Appl. **124**(5), 1741–1772 (2014)
8. Bean, N.G., O'Reilly, M.M., Palmowski, Z.: Matrix-analytic methods for the analysis of stochastic fluid-fluid models. Stoch. Model. **38**(3), 416–461 (2022)
9. Bright, L., Taylor, P.G.: Calculating the equilibrium distribution in level dependent quasi-birth-and-death processes. Stoch. Model. **11**(3), 497–525 (1995)
10. Buchholz, P., Mészáros, A., Telek, M.: Analysis of a two-state markov fluid model with 2 buffers. In: Iacono, M., Scarpa, M., Barbierato, E., Serrano, S., Cerotti, D., Longo, F. (eds.) Computer Performance Engineering and Stochastic Modelling - 19th European Workshop, EPEW 2023, and 27th International Conference, ASMTA 2023, Florence, Italy, June 20-23, 2023, Proceedings, vol. 14231, pp. 49–64. Springer, Cham (2023. https://doi.org/10.1007/978-3-031-43185-2_4
11. Buchholz, P., Mészáros, A., Telek, M.: Stationary analysis of a constrained markov fluid model with two buffers. Stochastic Models (2024). (to appear)
12. da Silva Soares, A., Latouche, G.: Matrix-analytic methods for fluid queues with finite buffers. Performance Eval. **63**(4-5), 295–314 (2006)
13. Hohn, N., Veitch, D., Papagiannaki, K., Diot, C.: Bridging router performance and queuing theory. SIGMETRICS Perform. Eval. Rev. **32**(1), 355–366 (2004)
14. Karandikar, R.L., Kulkarni, V.G.: Second-order fluid flow models: reflected Brownian motion in a random environment. Oper. Res. **43**, 77–88 (1995)
15. Latouche, G., Ramaswami, V.: Introduction to matrix analytic methods in stochastic modeling. SIAM (1999)
16. O'Reilly, M.M., Scheinhardt, W.: Stationary distributions for a class of Markov-modulated tandem fluid queues. Stoch. Model. **33**(4), 524–550 (2017)
17. Ramaswami, V.: Matrix analytic methods for stochastic fluid flows. In: International Teletraffic Congress, pp. 1019–1030, Edinburg (1999)
18. Stanford, D.A., Latouche, G., Woolford, D.G., Boychuk, D., Hunchak, A.: Erlangized fluid queues with application to uncontrolled fire perimeter. Stoch. Model. **21**(2–3), 631–642 (2005)

Stability of the Multiserver Job Queuing Model with Infinite Resources

Adityo Anggraito[1], Diletta Olliaro[1(✉)], Marco Ajmone Marsan[2], and Andrea Marin[1]

[1] Ca' Foscari University of Venice, Venice, Italy
{adityo.anggraito,diletta.olliaro,marin}@unive.it
[2] Institute IMDEA Networks, Leganés, Madrid, Spain
marco.ajmone@imdea.org

Abstract. A Multiserver Job Queuing Model (MJQM) is a queuing system that can be instrumental in the study of the dynamics of resource allocation in datacenters. The queue comprises a waiting line with infinite capacity and a large number of servers. In this paper, we look at the case of an infinite number of servers. Jobs are termed "multiserver" because each one is characterized by a resource demand in terms of number of simultaneously used servers and by a service duration. In a MJQM, jobs are clustered into classes, and a number of used servers is deterministically associated with each class. Instead, holding times are independent and identically distributed random variables whose distributions depend on the class of the job. We consider the case of just two job classes: "small" jobs use just one server, while "big" jobs use all servers in the system. The service discipline is First-Come-First-Served (FCFS). This means that if the job at the head-of-line (HOL) cannot enter service because the number of free servers is not sufficient to meet the job requirement, it blocks all subsequent jobs, even if there are sufficient free servers for them. Despite its importance, only few results exist for the MJQM, whose analysis is challenging, especially because the MJQM is not work-conserving. This implies that even the stability region of the MJQM is known only in special cases. In a previous work, we obtained a closed-form stability condition for MJQM with big and small jobs under the assumption of exponentially distributed service times for small jobs. In this paper, we compute the stability condition of MJQM with big and small jobs, with an infinite number of servers, considering different distributions of the service times of small jobs. Simulations are used to support the analytical results and to investigate the impact of service time distributions on the expected job waiting time before saturation.

Keywords: Multiserver Job Queuing Model · Stability region · Datacenter scheduler

1 Introduction

The interest in analytical tools that can improve the understanding of datacenter dynamics and thus guide the design of resource allocation algorithms with the

---

The original version of the chapter has been revised. Chapter title has been changed. A correction to this chapter can be found at
https://doi.org/10.1007/978-3-031-70753-7_11

A. Devos et al. (Eds.): ASMTA 2024, LNCS 14826, pp. 148–163, 2025.
https://doi.org/10.1007/978-3-031-70753-7_10

ultimate goal of achieving better performance has been growing together with the datacenter traffic, size and cost. The Multiserver Job Queuing Model (MJQM) [7] has emerged as one of the tools that can generate the insights necessary for the design of better algorithms. However, unfortunately, not many results are available, because of the difficulty in the analysis.

The MJQM is a queue with N identical servers and one FCFS waiting line with infinite capacity. Jobs arrive one by one according to a Poisson process with rate λ [s^{-1}]. Jobs are termed "multiserver" because each one is characterized by a resource demand in terms of number of simultaneously used servers and by a random service duration. In a MJQM, jobs are clustered into classes, and a number of used servers is deterministically associated with each class. We consider the case of just two job classes: "small" jobs use just one server, while "big" jobs use all servers in the system. We denote by p_s the probability that an arriving job is small and by $p_b = 1 - p_s$ the probability that an arriving job is big. The holding times of the required number of servers are independent and identically distributed random variables whose distributions depend on the class of the job. We take our unit of time to be seconds and we indicate it as [s]. We denote by $1/\mu_s$ [s] the average server holding time of small jobs and by $1/\mu_b$ [s] the average server holding time of big jobs. The FCFS service discipline is such that if the job at the head-of-line (HOL) cannot enter service because the number of free servers is not sufficient to meet the job requirement, it blocks all subsequent jobs, even if there are sufficient free servers for them.

The total workload of the queue is $\rho = \lambda p_s/\mu_s + N\lambda p_b/\mu_b$. The queue is not work-conserving because of HOL blocking. As a result, the queue stability condition is not just $\rho < N$.

In a previous work [9], we computed the stability condition of a MJQM with big and small jobs under the assumption of exponentially distributed service times for small jobs, also proving the insensitivity of the result to the service time distribution of big jobs. In this paper, we extend the analysis to several service time distributions of small jobs under the assumption of an infinite number of servers. It is worth noting that real public datacenters comprise very large numbers of machines, so that the study of the case of an infinite number of servers can provide interesting insights. The service time distributions that we consider for small jobs are uniform, exponential, and Fréchet, in addition to the case of constant service times. The constant and uniform cases are representative of variances smaller than the exponential case. The Fréchet distribution is often used in extreme value statistics, and exhibits a heavy tail. Our analysis confirms the insensitivity of the stability region to the distribution of big jobs.

The major findings of this paper can be summarized as follows:

– Differently from other well-known queuing systems (e.g., G/G/k) the stability region of the model depends on moments higher than the first of the distribution of small jobs' service times. This result answers to the question remained open in [9], where we proved the insensitivity of the stability condition to the distribution of big job durations but left the problem of deciding the (in)sensitivity for the small ones unanswered.

– We provide an exact stability analysis for some special distributions: deterministic, uniform and Fréchet, as well as the exponential distribution already considered in [9].
– We show that for a large class of distributions (those with slow varying complementary cumulative density function), we can give a stability condition that is asymptotically correct and, for practical scenarios, provides accurate estimates of the maximum arrival rate. We apply this method to approximate the stability condition for Pareto distributed service times.
– We study by simulation the impact of the small jobs' tail distribution on the system expected waiting times in the light of the previous results.

The paper is structured as follows. Section 2 reviews the state of the art. In Sect. 3, we show that MJQM is sensitive to the distribution of the service time of small jobs and provide the exact stability conditions for some important distributions. In Sect. 4, we illustrate a method based on the Fisher-Tippet theorem to estimate the stability condition for a large class of heavy-tailed distributions. Then, in Sect. 5, we show some discrete event simulation results on the jobs' average waiting time. Finally, Sect. 6 concludes the work.

2   Related Work

The stability region for the MJQM has long puzzled researchers, posing a challenge to the deep understanding of this particular model. Despite researchers efforts, results have been limited to corner cases and approximations. Outcomes presented in [4] and [3] are restrained to scenarios in which the system is composed of 2 servers. The authors of [1] studied the system for an arbitrary number of servers and derived the stability condition under the assumption that all jobs have the *same* exponential service time distribution. This result has been later generalized in [8] admitting more general arrival processes. The contribution in [2] has been instrumental in addressing this problem, thanks to the *saturation rule*, which considers systems having always an infinite amount of waiting jobs. Recently, a simple closed-form expression for the two-classes case under the assumption that jobs have different exponential service times has been obtained in [5].

Advancing further on this trail, in [9], we have derived an explicit stability condition expression in terms of the incomplete Euler Beta function which can be computed efficiently at arbitrary precision and does not require the generation of the whole state space. Moreover, the analysis we have proposed, based on renewal theory, is remarkably insensitive to the service time distribution of big jobs. While these findings have contributed significantly to the unveiling of some MJQM dynamics, the stability region for this particular system still dodges complete understanding. Regrettably, scant attention has been paid to the impact of service time distributions for the class of small jobs. To address this critical research gap, we leverage another interesting aspect proposed in [9]: the asymptotic analysis for the case in which the number of servers, N, tends to infinity, which is going to be the starting point for the analysis we propose here.

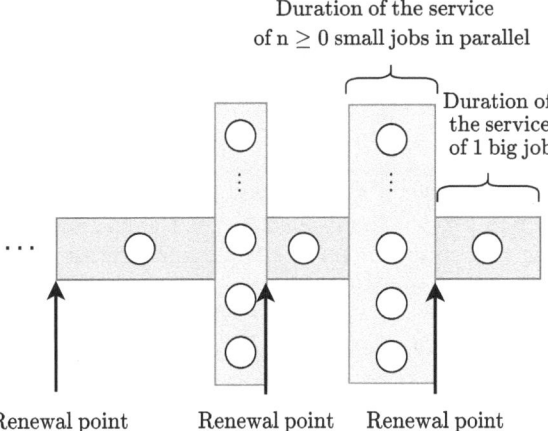

Fig. 1. Renewal cycle of the stochastic process underlying the MJQM with small and big jobs.

3   Sensitivity to Small Jobs' Duration

In this section, we study the impact on the stability condition of the service time distribution of small jobs in a MJQM with infinite servers.

We start by observing that in work-conserving queuing systems like, e.g., the well-known GI/G/k queue, the distribution of the service times have an impact on the system performance indices but not on the stability region. If λ and μ are the arrival and service rates of a GI/G/k system, respectively, the stability condition is $\lambda < k\mu$, independently of the further moments of the service time.

In a MJQM where big jobs take all available servers, we know that the stability condition depends on the average service time of big jobs only, not on their distribution. In this section, we explain why this does not hold for the small jobs' service times. We consider the following simplified scenario:

- Small jobs require just one server
- Big jobs require all servers
- The number of servers is infinite

In order to derive the stability condition, we follow the reasoning of [2,5,9] and consider the maximum throughput of a saturated system, i.e., a MJQM whose waiting line always contains an infinite number of jobs waiting to be served.

3.1   General Case

We consider the completion of a service of a big job as a renewal point of the stochastic process underlying the saturated MJQM system. The renewal cycle comprises the service of a sequence of n small jobs, where $n \geq 0$, and exactly one big job. The average service time of a big job is μ_b^{-1} and that is also the duration of the phase depicted in red in Fig. 1. Notice that, due to the Poisson assumption, n is a geometrically distributed random variable with parameter p_s, i.e.:

$$Pr\{n = k\} = p_s^k p_b.$$

Since we have infinite servers, all the small jobs in the waiting line before the next big job enter in service simultaneously at the end of service of a big job. The following big job enters in service when the slowest small job completes service. Let $F_s(t)$ be the cumulative distribution function (CDF) of the service time distribution of small jobs, and let $f_s(t)$ be the corresponding probability density function (pdf). Let $X_1, \ldots, X_n$ be the independent random variables distributed according to $F_s(t)$ modelling the service duration of a small job. Let us define the maximum order statistics as:

$$X_{n:n} = \max_{i=1,\ldots,n} (X_i).$$

The distribution of the duration of the phase of service of small jobs is the same of $X_{n:n}$ whose pdf is:

$$f_{n:n}(t) = n[F_s(t)]^{n-1} f_s(t), \qquad n > 0.$$

The average duration of a small job service phase T_s given the number of small jobs is:

$$E[T_s|n] = \begin{cases} 0 & \text{if } n = 0 \\ \int_0^\infty t f_{n:n}(t)dt & \text{if } n > 0 \end{cases}, \tag{1}$$

Since n is geometrically distributed, we can decondition this mean as follows:

$$E[T_s] = \sum_{n=1}^\infty E[T_s|n]p_s^n p_b. \tag{2}$$

Therefore, the average length $E[T_{bb}]$ of a renewal cycle is:

$$E[T_{bb}] = \frac{1}{\mu_b} + p_b \sum_{n=1}^\infty p_s^n \int_0^\infty nt[F_s(t)]^{n-1} f_s(t)dt. \tag{3}$$

Notice that the rewriting of the improper integral and of the infinite summation in closed-form may be challenging. Hereafter, we consider some distributions that admit closed-form expressions for the stability conditions, some with low tail and one with heavy tail, as well as the exponential case.

Following [9], the stability condition obtained as maximum throughput of the saturated model is:

$$\lambda p_b E[T_{bb}] < 1. \tag{4}$$

Note that the average duration of the renewal cycle is finite, even if the MJQM is saturated, provided the average duration of both service phases is finite.

3.2 Exponential Service Times for Small Jobs

The case of exponentially distributed service times for small jobs was considered in [9] in the case of a finite number of servers. The case of infinite servers was studied as a limiting case. Since the approach that we use here is different and it is based on the maximum order statistics, we reproduce the results in [9] with this new tool as a sanity check.

We can rewrite Eq. (1) as:

$$E[T_s^{\exp}|n] = \frac{H_n}{\mu_s},$$

where H_n is the n-th harmonic number. Hence, we derive:

$$E[T_s^{\exp}] = -\frac{\log(1 - p_s)}{\mu_s} = -\frac{\log(p_b)}{\mu_s}.$$

In conclusion, we have:

$$E[T_{bb}^{\exp}] = \frac{1}{\mu_b} - \frac{\log(p_b)}{\mu_s}, \tag{5}$$

reproducing the results of [9, Eq. (8)], as expected.

3.3 Deterministic and Uniform Service Times for Small Jobs

The maximum order statistics of independent uniform random variables has been known for a long time. Consider a uniform random variable in the interval (a, b) with $0 \le a < b$, with $a = b$ becoming the deterministic degeneration of the random variable. We can rewrite Eq. (1) for $n > 0$ as:

$$E[T_s^{uni}|n] = \frac{a + bn}{n + 1}.$$

The following proposition gives $E[T_s^{uni}|n]$.

Proposition 1. *The average duration of a phase of service of small jobs for uniformly distributed service times is given by:*

$$E[T_s^{uni}] = (b - a)p_b \left(1 + \frac{\log(p_b)}{p_s}\right) + bp_s. \tag{6}$$

Proof. By Eq. (2), we have to evaluate the series:

$$\sum_{n=1}^{\infty} \frac{a + bn}{n + 1} p_s^n p_b = p_b \left((a - b) \sum_{n=1}^{\infty} \frac{p_s^n}{n + 1} + b \frac{p_s}{p_b}\right)$$

$$= p_b \left(\frac{a - b}{p_s} \sum_{n=1}^{\infty} \int_0^p x^n dx + b \frac{p_s}{p_b}\right)$$

$$= p_b \left(\frac{b - a}{p_s}(p_s + \log(p_b)) + b \frac{p_s}{p_b}\right),$$

which can be easily simplified to obtain Eq. (6). $\qquad\qquad\square$

In conclusion, for the uniform distribution of the small job duration, we can write:

$$E[T_{bb}^{\text{uni}}] = \frac{1}{\mu_b} + (b-a)p_b \left(1 + \frac{\log(p_b)}{p_s}\right) + bp_s \,,$$

where, if $a = b$, we get the deterministic case expression.

3.4   Fréchet Distributed Service Times for Small Jobs

The Fréchet distribution, or inverse Weibull distribution, is particularly useful to study max order statistics. It is a distribution that belongs to the class of generalized extreme value (GEV) distribution. Beside its intrinsic interest since it exhibits a heavier tail than the exponential distribution, the Fréchet distribution can be often used to approximate the maximum of a set of independent random variables under appropriate conditions [6].

The CDF of the distribution with support $(0, \infty)$ is:

$$F_F(t) \triangleq Pr\{F \le t\} = \exp\left(-(x/\xi)^{-\alpha}\right) \,,$$

where $\alpha > 0$ and $\xi > 0$ are the shape and scale parameters, respectively. The Fréchet distribution has finite mean under the condition $\alpha > 1$ and finite variance for $\alpha > 2$ and is heavy-tailed.

The Fréchet distribution enjoys some important properties. The first that we present is connected to its max order statistics and is known as the maximum stability postulate. Let $X_1, \ldots, X_n$ be i.i.d. Fréchet random variables with parameters α and ξ. Then, $X_{n:n}$ has a Fréchet distribution with parameters α and $n^{1/\alpha}\xi$, i.e., the shape parameter of the maximum is the same of the individual random variables, while the scale parameter depends on n. Therefore, the expectation of the maximum order statistics of n i.i.d. random variables distributed according to Fréchet law with parameters ξ, α is:

$$E[T_s^{\text{fré}}|n] = -\xi\, n^{1/\alpha} \Gamma\left(\frac{1-\alpha}{\alpha}\right) \,,$$

where Γ is the Euler gamma function.

Another important property of the Fréchet distribution will be used in Sect. 4: under certain conditions, the max order statistics of heavy-tailed distributions converge, under appropriate normalization, to a Fréchet distribution. In other words, if the service time distribution of the small jobs is Fréchet distributed, the results presented in this section are exact, otherwise, under mild conditions, they represent an approximation that converges when p_s gets close to 1.

Now, we are in position to decondition the mean.

Proposition 2. *The average duration of a phase of service of small jobs for Fréchet distributed service times is given by:*

$$E[T_s^{\text{fré}}] = p_b\, \xi\, \Gamma\left(\frac{\alpha-1}{\alpha}\right) Li_{-1/\alpha}(p_s)\,, \tag{7}$$

where $Li_{-1/\alpha}(p_s)$ is the polylogarithm function, i.e., in our case:

$$Li_{-1/\alpha}(p_s) = \sum_{k=1}^{\infty} p_s^k k^{1/\alpha}.$$

Proof. The result is trivial given the expression of $E[T_s^{\mathrm{fré}}|n]$.

In conclusion, we have:

$$E[T_{bb}^{\mathrm{fré}}] = \frac{1}{\mu_b} + p_b\, \xi\, \Gamma\left(\frac{\alpha-1}{\alpha}\right) \mathrm{Li}_{-1/\alpha}(p_s). \tag{8}$$

Recall that the polylogarithmic function can be efficiently computed at arbitrary precision and is implemented in most of the mathematical software.

3.5   Is Variance Everything We Need to Know?

It may be intuitive to formulate a conjecture stating that higher variances of the service time distribution for small jobs lead to more restrictive stability conditions. Quite surprisingly, this is untrue and we provide in this section a numerical counterexample.

Let us consider a MJQM system S_1 with the following parameters: $p_s = 0.9$, $\mu_b = 1$, $\mu_s = 1$ where all times are exponentially distributed. By using Eq. (5), we can readily obtain the stability condition for the intensity of the arrival stream which is $\lambda < 3.02793$. Consider now a system S_2 with Fréchet distributed service times for the small jobs with parameters $\xi = 0.654147$ and $\alpha = 2.4$. These parameters imply a unitary average service time. In this case, we use Eq. (8) and obtain $\lambda < 3.1998$, i.e., the condition is less restrictive than the one of the exponential distribution (system S_1). Hence, the stability region of S_2 is wider than the one of S_1. However, the coefficient of variation of system S_2 is 1.18, i.e., higher than the one of the exponential distribution (system S_1). We thus have that a higher service time variance produces a wider stability region.

Why does this happen? To understand this phenomenon, we can consult Fig. 2. While the heavy tail of the Fréchet distribution is visible from the plot of Fig. 2, it is possible to see that the distribution concentrates a high probability mass around its mode. If the probability of long sequences of jobs is sufficiently small, the adverse event of a very long job is compensated by the appearance of many jobs with similar service times. Therefore, even if the variance is higher for the Fréchet, the mixture of big and small jobs makes the stability region of the exponential service times smaller than the Fréchet's one.

3.6   Comparison of Stability Regions

We start by examining the impact of the value of the α parameter of the Fréchet distribution on the stability region of the MJQM. In Fig. 3 we show curves of λ_{max}, the upper extreme of the stability region (the lower extreme obviously is always zero) as a function of α, for different values of p_b (left chart) or μ_b (right chart). We can see that, assuming $\alpha > 1$ in order to obtain a finite mean, increasing values of α correspond to wider stability regions, with a diminishing return effect. It must be noted that for values $\alpha \le 2$ the variance of the Fréchet distribution is infinite. For these reasons, in the next graphs we consider both cases of $1 < \alpha \le 2$ and $\alpha > 2$.

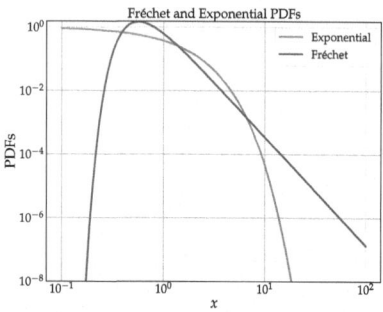

Fig. 2. pdf of the exponential and Fréchet distributions in log-log-scale with mean 1.

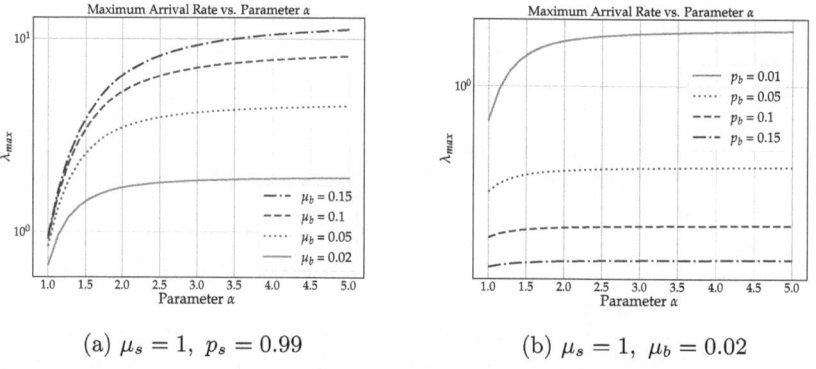

(a) $\mu_s = 1,\ p_s = 0.99$ (b) $\mu_s = 1,\ \mu_b = 0.02$

Fig. 3. Maximum arrival rate vs. Fréchet's α parameter.

Figure 4 plots curves of the upper extreme of the stability region for the four considered service times distributions of small jobs (deterministic, uniform, exponential and Fréchet) versus the service rate of small jobs μ_s for two values of p_b. In the case of the Fréchet distribution we consider $\alpha = 1.5$ as well as $\alpha = 2.15$. For both values of p_b we see that the stability region becomes wider for shorter average service time of small jobs, due to the reduced influence of the maximum small job service time with respect to the big job service time. In addition, we observe that with $p_b = 0.01$ the values of λ_{max} are almost one order of magnitude larger. This is due to the dependency of λ_{max} on $1/p_b$ in Eq. (4). The ranking of the widths of the stability regions is as expected: the widest stability region is obtained with deterministic service times of small jobs, followed by uniform, exponential, Fréchet with finite variance and finally the smallest stability region is obtained with the Fréchet with infinite variance. With $p_b = 0.1$ the stability region of the Fréchet with finite variance is very close to the exponential, but the difference grows larger for $p_b = 0.01$.

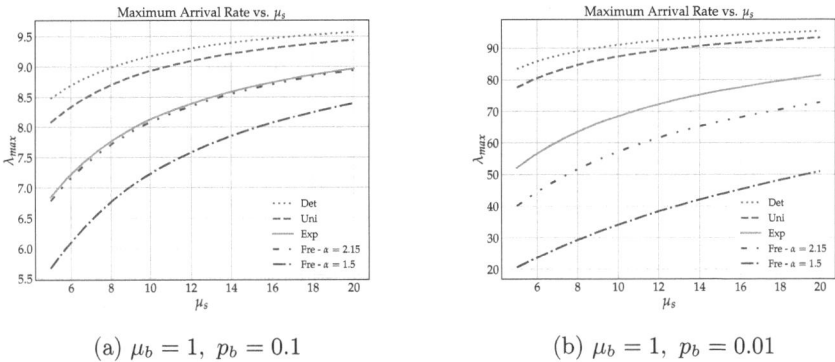

(a) $\mu_b = 1$, $p_b = 0.1$ (b) $\mu_b = 1$, $p_b = 0.01$

Fig. 4. Maximum arrival rate vs. μ_s.

Finally, in Fig. 5 we present the curves of λ_{max} versus p_b for the different service time distributions of small jobs and for average small jobs service times equal to either 1 (left chart) or 0.1 (right chart), assuming an average service time of big jobs equal to 10. As expected, when the difference between big and small job service times reaches two orders of magnitude, the pdf of small jobs service times has a limited impact on the stability region.

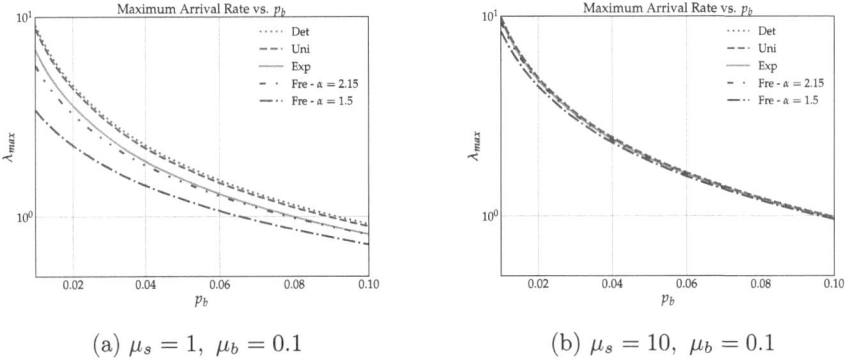

(a) $\mu_s = 1$, $\mu_b = 0.1$ (b) $\mu_s = 10$, $\mu_b = 0.1$

Fig. 5. Maximum arrival rate vs. p_b.

4   Asymptotic Approximations

In this section, we use results of extreme value theory to derive an approximate, asymptotically correct, stability condition for the MJQM with infinite servers and two classes of jobs, small and big.

Suppose that $p_s \to 1^-$, i.e., we are sure that the sequence of small jobs waiting in a line between two big jobs is sufficiently long. Notice that this is an assumption usually met in real systems, where jobs with low service demand appear more frequently than big ones [12]. For instance, what emerged from the analysis of eight cells from a Google Borg datacenter, considering only classes in which jobs fall with reasonably high probability and normalising these probabilities again, is that single-threaded jobs compose a range of the total workload between 83% and 96% [12] and the probability of the largest-sized jobs is of the order of 10^{-5}. Grouping jobs in large and small thus leads to values of p_s not far from 1.

If the length of the sequence of small jobs is sufficiently long, we can apply the Fischer-Tippet theorem (see, e.g., [10, 11]).

Result 1 (Fisher-Tippet Theorem) *Let $X_1, \ldots, X_n$ be i.i.d. random variables with CDF $F(x)$. We call $X_{n:n} = \max(X_i)$. Let (a_n) and (b_n) be constants such that for some nondegenerate limit distribution G we have:*

$$\lim_{n \to \infty} Pr\{\frac{X_{n:n} - b_n}{a_n} \le x\} = G(x), \quad x \in \mathbb{R}.$$

Then, $G(x)$ must have one of these three possible forms:

Type 1: *Fréchet distribution with parameters $\alpha > 0$ and $\xi = 1$;*
Type 2: *Weibull distribution with parameter $\alpha > 0$;*
Type 3: *Gumbel distribution.*

Informally, Result 1 states that if we are able to find the normalizing constants a_n, b_n, and the limit of the CDF of the maximum converges to a nondegenerate CDF $G(x)$ as $n \to \infty$, then $G(x)$ can have only three possible shapes. Clearly, the difficult part is to determine how we get the parameter of the Fréchet or Weibull distributions, and how to find the normalizing constants a_n and b_n.

We say that a CDF F belongs to the max-domain of attraction of G if the distribution function of the normalized maximum converges to G. In particular, we are interested in characterizing the distributions whose maximum asymptotically tends to the Fréchet distribution, i.e., we are looking for the maximum domain of attraction (MDA) of the three types of distributions.

Result 2 (MDA of the Fréchet distribution [11]) *A CDF F belongs to the domain of attraction of the Fréchet distribution with parameter $\alpha > 0$ if and only if function $1 - F(x)$ is regularly varying with index α, i.e.:*

$$\lim_{t \to \infty} \frac{1 - F(tx)}{1 - F(t)} = x^{-\alpha}.$$

Quite interestingly, it can be shown that if F is the distribution of r.v. X and it belongs to the MDA of the Fréchet distribution, then $E[X^K] \to \infty$ for $k > \alpha$.

Result 2 allows us to study a wide range of heavy-tailed distribution and, in particular, the Pareto distribution, which has shown to be quite useful in modeling job durations [7].

Consider a Pareto distribution with parameters x_m and α:

$$Pr\{X > x\} = \begin{cases} \left(\frac{x_m}{x}\right)^\alpha & x \geq x_m \\ 1 & \text{otherwise} \end{cases}.$$

The normalizing constants for the limiting Fréchet distribution are $a_n = x_m n^{1/\alpha}$ and $b_n = 0$, while the shape parameter of the Fréchet corresponds to the shape parameter of the Pareto [11]. Now, we can estimate the length of T_s^{par} when p_s is large, i.e.:

$$E[T_s^{\text{par}}] \simeq x_m p_b \Gamma\left(1 - \frac{1}{\alpha}\right) Li_{-1/\alpha}(p_s).$$

To have an idea of the approximation on the average duration of the phase of service of small jobs, we show some numerical evidences in Table 1. We notice that for $p_s = 0.9$, i.e., when the expected length of the sequence of small jobs is 10, then $E[T_s^{\text{par}}]$ falls outside the confidence interval of the simulation, but the accuracy is already good (with a relative error lower than 3%). However, for longer sequences we may see that the approximation based on the Fréchet distribution is close to the center of the confidence interval.

Table 1. Simulation of $E[T_s^{\text{par}}]$ with a Pareto distribution with parameter $x_m = 1$ and $\alpha = 1.5$ for different values of p_s. We consider 30 independent experiments, each of which with 10^6 sequences of small jobs. The interval is constructed with a confidence of 95%.

p_s	$E[T_s^{\text{par}}]$ sim.	$E[T_s^{\text{par}}]$ approx.
0.9	$(10.5034, 10.5749)$	10.2484
0.99	$(50.6637, 52.2755)$	51.6645
0.999	$(239.0902, 242.3966)$	241.638

5   Impact of the Service Time Distribution on the Expected Waiting Times

While the analytical results based on the max order statistics and its asymptotic behaviour are useful to determine the stability region of a MJQM system, in order to study the impact of the distribution of small/big jobs on performance indices, we need to resort to a discrete event simulator. Recall that the insensitivity property of the duration of the big jobs holds for the characterization of the stability region, while the performance measures still depend at least on its second moment. To see this, consider a MJQM where $p_s \to 0^+$, and observe that the system tends to behave as a M/G/1 queue, and it is well-known that the P-K formula for the expected waiting time depends on the second moment of the distribution of the (big) jobs.

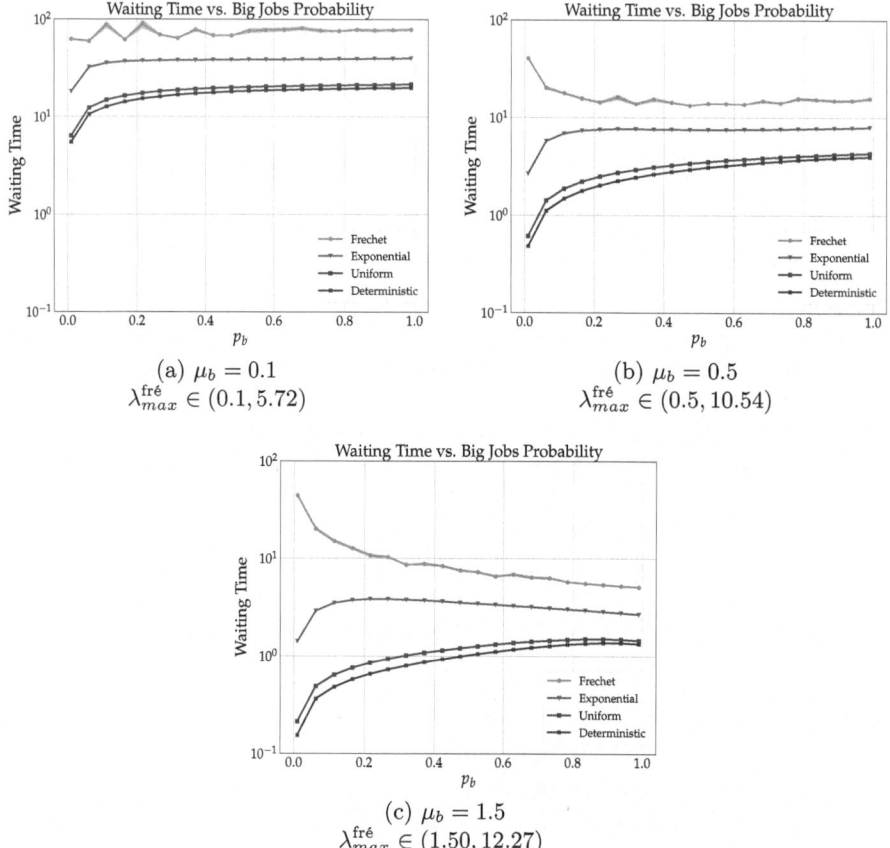

Fig. 6. Overall expected waiting time vs. big jobs probability; $\mu_s = 1$, $\alpha = 2.15$. λ varies according to p_b and is always taken as the 80% of the $\lambda_{max}^{fré}$.

Impact of Frequency and Duration of Small and Big Jobs on the Expected Waiting Time. In the first simulation (Fig. 6), we study the overall expected waiting time (of both big and small jobs) as a function of the frequency and duration of big jobs. The results are produced considering 15 independent experiments and 50 million events for each experiment, with the exception of the experiments involving the Fréchet distribution that required 100 million events to account for its heavy tail. We consider four distributions for the duration of the small and big jobs: exponential, uniform, deterministic and Fréchet. Notice that for the latter distribution, to estimate the mean waiting time, we need to assume $\alpha > 2$. The service rate of big jobs is 10% (Fig. 6a), 50% (Fig. 6b) and 150% (Fig. 6c) the service rate of small jobs. In the scenarios considered here, big and small jobs service times are always sampled from the same distribution. It is interesting to see that in Figs. 6b and 6c, the expected waiting time of the Fréchet has a decreasing trend while all other distributions exhibit a growing trend at least for low values of p_b. The explanation is intriguing. As p_b grows, the distributions with low variance show higher waiting times as intuitively expected since big jobs (even when they are faster than small ones) require all servers. However, the Fréchet distribution is heavy-tailed, and when there is a sufficiently large number of small jobs between two big ones, it becomes more likely that the maximum can be very long and thus the waiting time is increased by the waste of resources. Long sequences of small jobs correspond to small values of p_b. Therefore, while for distributions without heavy-tail the first effect is prominent, hence small values of p_b imply a smaller expected waiting time, for heavy-tailed distributions the second effect is prominent. In fact, we can see that when $\mu_b = 0.1$, i.e., big jobs are slow and hence consume more resources, the initial decreasing phase of the Fréchet is not visible any more, because the benefits of reducing p_b are roughly equivalent to the negative effects caused by increasing the length of the sequences of small jobs.

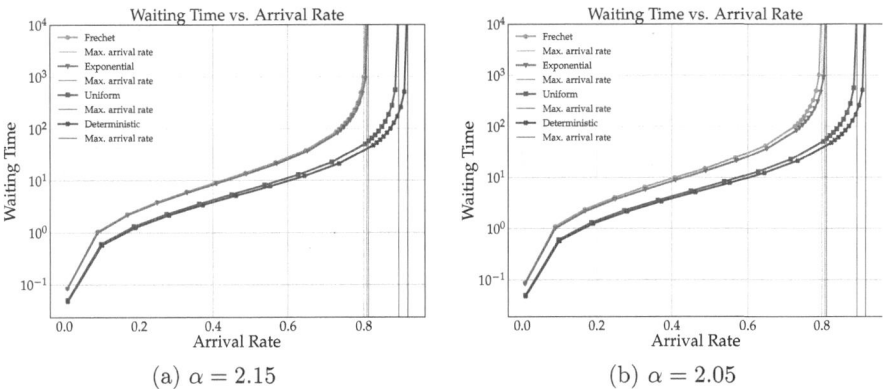

Fig. 7. Expected waiting time vs. arrival rate; $\mu_s = 1$, $\mu_b = 0.1$, $p_b = 0.1$

Expected Waiting Time As Function of the Intensity of the Arrival Process. In the second simulation study (Figs. 7 and 8), we study the overall expected waiting time, with increasing arrival rate. In the plots of this section, we show the vertical asymptotes denoting the stability region of the MJQM obtained with the

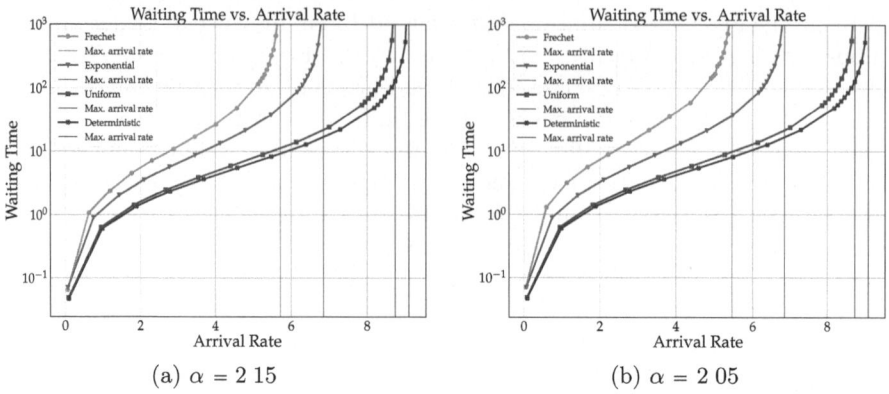

Fig. 8. Expected waiting time vs. arrival rate; $\mu_s = 1$, $\mu_b = 0.1$, $p_b = 0.01$

formulas of Sect. 3. Notice that in this set of experiments, to minimize noise in the results of Fréchet distribution, big jobs are exponentially distributed instead. Figures 7a and 7b show very similar performance between the exponentially and Fréchet distributed services times. In addition, as expected, uniform and deterministic distributions perform better due to their low variability. We also see that in the case of Fréchet distribution, lowering the α value, which means increasing the impact of a heavier tail, does not give significant impact to the stability point. To understand this surprising result, we need to observe the results of Figs. 8a and 8b. These figures show that as we increase the frequency of small jobs (increasing p_s from 0.9 to 0.99), the waiting time and the stability region get worse. Hence, the stability regions of the Fréchet distribution are significantly lower than those associated with the exponential distribution. Notice also the impact of the α parameter in the Fréchet. As it decreases, the stability region of the system becomes smaller since (i) the heavy-tail of the distribution plays an important role and (ii) the higher presence of long sequences of small jobs which increase the probability of having long small jobs in the system.

6 Conclusion

With the objective of contributing analytical results that can increase the understanding of job service dynamics in datacenters, in this paper we have investigated the stability condition of multiserver job queuing models with an infinite number of servers and two classes of customers: big and small jobs.

We have computed explicit expressions for the stability region of systems where service times of small jobs have different probability distributions, and big jobs can have any service time pdf.

We have shown that the stability region width depends on the pdf of small job service times, in contrast to what happens in traditional queues where the stability region only depends on average service times.

Additionally, we have also shown that, contrary to what could be expected, the width of the stability region does not always grow for decreasing service time variance. This implies a dependence on higher moments of the service time distribution that needs to be further investigated.

References

1. Afanaseva, L., Bashtova, E., Grishunina, S.: Stability analysis of a multi-server model with simultaneous service and a regenerative input flow. Methodol. Comput. Appl. Probab. **22**(4), 1439–1455 (2020)
2. Baccelli, F., Foss, S.: On the saturation rule for the stability of queues. J. Appl. Probab. **32**(2), 494–507 (1995)
3. Brill, P., Green, L.: Queues in which Customers Receive Simultaneous Service from a Random Number of Servers: a System Point Approach. Manage. Sci. **30**(1), 51–68 (1984)
4. Filippopoulos, D., Karatza, H.: An M/M/2 parallel system model with pure space sharing among rigid jobs. Math. Comput. Modelling **45**(5), 491–530 (2007)
5. Grosof, I., Harchol-Balter, M., Scheller-Wolf, A.: New stability results for multiserver-job models via product-form saturated systems. SIGMETRICS Performance Eval. Rev. **51**(2), 6–8 (2023)
6. Haan, L., Ferreira, A.: Extreme value theory: an introduction. Springer (2007)
7. Harchol-Balter, M., Downey, A.B.: Exploiting process lifetime distributions for dynamic load balancing. ACM Trans. Comput. Syst. **15**(3), 253–285 (1997)
8. Morozov, E., Rumyantsev, A.: Stability analysis of a $MAP/M/s$ cluster model by matrix-analytic method. In: Fiems, D., Paolieri, M., Platis, A.N. (eds.) EPEW 2016. LNCS, vol. 9951, pp. 63–76. Springer, Cham (2016). https://doi.org/10.1007/978-3-319-46433-6_5
9. Olliaro, D., Ajmone Marsan, M., Balsamo, S., Marin, A.: The saturated multiserver job queuing model with two classes of jobs: Exact and approximate results. Perform. Eval. **162**, 102370 (2023)
10. Paul Embrechts, S.I.R., Samorodnitsky, G.: Extreme value theory as a risk management tool. North Am. Actuarial J. **3**(2), 30–41 (1999)
11. Roncalli, T.: Handbook of Financial Risk Management. Chapman and Hall/CRC (2020)
12. Tirmazi, M., et al.: Borg: the next generation. In: Proceedings of the Fifteenth European Conference on Computer Systems, pp. 30:1–30:14. Association for Computing Machinery (2020)

Correction to: Stability of the Multiserver Job Queuing Model with Infinite Resources

Adityo Anggraito[ORCID], Diletta Olliaro[ORCID], Marco Ajmone Marsan[ORCID], and Andrea Marin[ORCID]

Correction to:
Chapter 10 in: A. Devos et al. (Eds.): *Analytical and Stochastic Modelling Techniques and Applications*, LNCS 14826, https://doi.org/10.1007/978-3-031-70753-7_10

The original version of chapter 10 title was inadvertently captured. By revising the title, this has been corrected.

The updated version of this chapter can be found at
https://doi.org/10.1007/978-3-031-70753-7_10

Author Index